# NUCLEAR COPPER

# NUCLEAR COPPER

## THE SECRET WORLD OF NUCLEAR POLICING

MATT OKUHARA

AMBERLEY

First published 2025

Amberley Publishing
The Hill, Stroud
Gloucestershire, GL5 4EP

www.amberley-books.com

Copyright © Matt Okuhara, 2025

The right of Matt Okuhara to be identified as the Author of this work has been asserted in accordance with the Copyright, Designs and Patents Act 1988.

ISBN 978 1 3981 2487 5 (hardback)
ISBN 978 1 3981 2488 2 (ebook)

All rights reserved. No part of this book may be reprinted or reproduced or utilised in any form or by any electronic, mechanical or other means, now known or hereafter invented, including photocopying and recording, or in any information storage or retrieval system, without the permission in writing from the Publishers.

British Library Cataloguing in Publication Data.
A catalogue record for this book is available from the British Library.

1 2 3 4 5 6 7 8 9 10

Typesetting by SJmagic DESIGN SERVICES, India.
Printed in the UK.

Appointed GPSR EU Representative: Easy Access System Europe Oü, 16879218
Address: Mustamäe tee 50, 10621, Tallinn, Estonia
Contact Details: gpsr.requests@easproject.com, +358 40 500 3575

# AUTHOR'S NOTE

For security and confidentiality reasons, certain details such as names, places, and dates have been altered in this book. Drawing from personal diary entries, recollections, and discussions with fellow officers during my tenure with the Civil Nuclear Constabulary (CNC), I have endeavoured to faithfully reconstruct conversations and events to the best of my ability. Mindful of the sensitive nature of nuclear security, great care has been taken to avoid any disclosure of confidential information that could jeopardise ongoing or future operations.

To provide readers with an insight into the work of a CNC officer while safeguarding operational tactics and strategies, certain incidents and activities have been amalgamated to present a broader depiction of their responsibilities. Additionally, references to events have been cross-checked against publicly available records obtained through freedom of information requests, public document searches, and contemporary news reports.

This memoir serves as a firsthand account of the training, duties, and challenges faced within the CNC, shedding light on the dedicated efforts of law-enforcement professionals across the UK. I extend my appreciation to the entire law-enforcement community, with particular gratitude to my colleagues in the Civil Nuclear Constabulary for their unwavering commitment to duty.

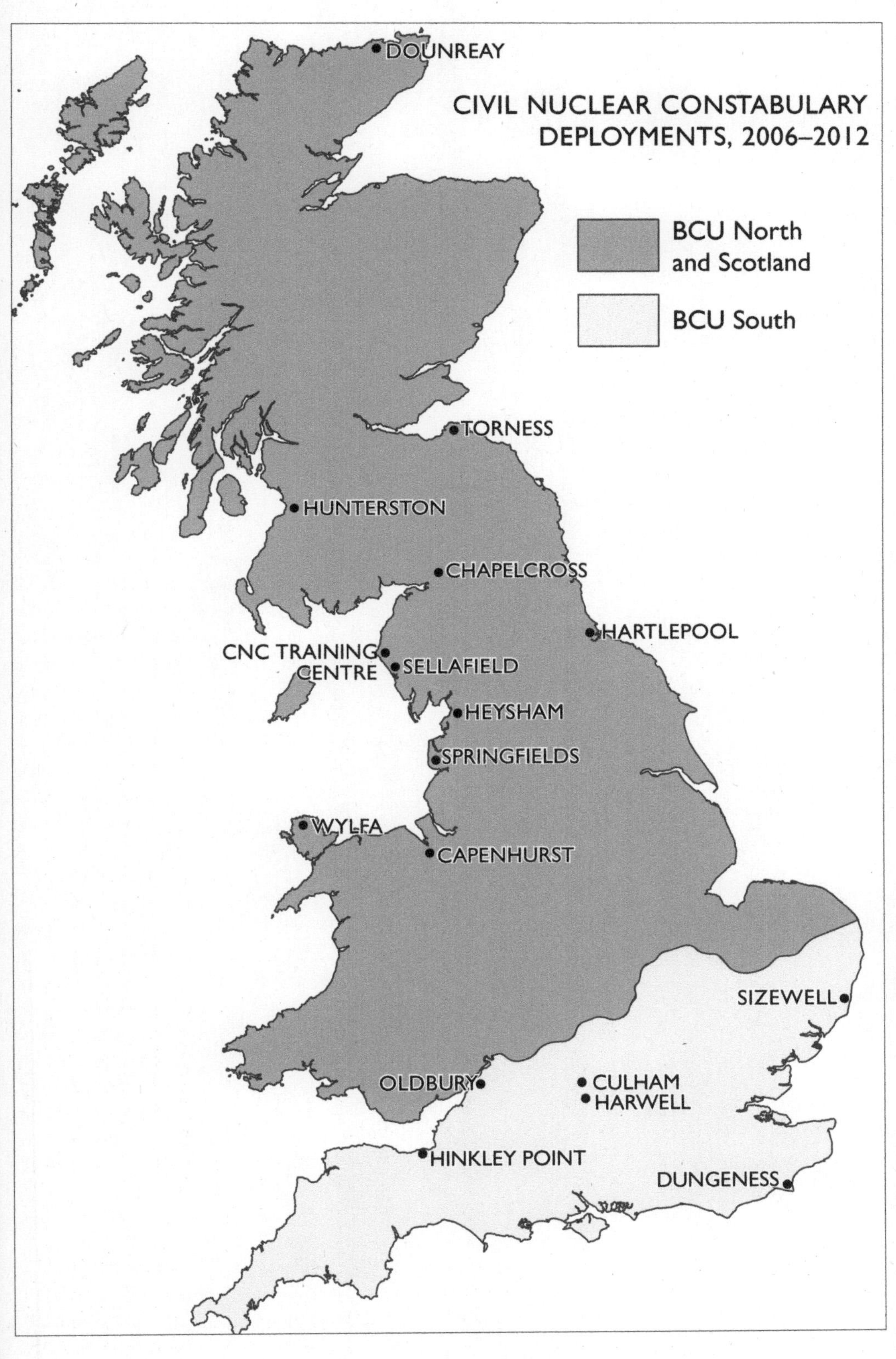
CIVIL NUCLEAR CONSTABULARY
DEPLOYMENTS, 2006–2012
BCU North
and Scotland
BCU South
DOUNREAY
TORNESS
HUNTERSTON
CHAPELCROSS
HARTLEPOOL
CNC TRAINING
CENTRE
SELLAFIELD
HEYSHAM
SPRINGFIELDS
WYLFA
CAPENHURST
SIZEWELL
OLDBURY
CULHAM
HARWELL
HINKLEY POINT
DUNGENESS

# LIST OF ILLUSTRATIONS

of the same techniques employed by the officers that make up the UK police service. (Crown Copyright/National Archives)

9. M134 minigun – used for defence aboard the escort vessels. (Crown Copyright/National Archives)
10. General-purpose machine gun (GMPG), mounted for defence. (Crown Copyright/National Archives)
11. The *Pacific Heron* and *Pacific Egret*. Both of these vessels have been used to move MOX fuel around the world. (Courtesy of PNTL Ltd)
12. Loading the fuel. Using specially constructed containers, the loading process can take several hours to complete. (Courtesy of PNTL Ltd)
13. Counter-terrorism exercises. Officers regularly train with other security agencies and militaries. (Crown Copyright)
14. Road escort officers conducting a briefing in between training exercises. (Author's collection)
15. A C130 Hercules at Wick Airport, similar to the type used to carry out a short-notice cargo move to the USA. (Courtesy of Highland and Island Airports Ltd)
16. Firefighter training as part of the escort group pre-embarkation package. (Author's collection)
17. The *Atlantic Osprey* arriving in German territorial waters, escorted by the *Wasserschutzpolizei*. (Courtesy of Fred Dott/Greenpeace Media)
18. The *Atlantic Osprey* coming into Nordenham, with German escort and protesters surrounding the vessel. (Courtesy of DPA Picture Alliance Archive/Alamy)
19. Manning a GPMG on the *Atlantic Osprey*. (Author's collection)
20. Having operated from 1955 until 2005, the United Kingdom Atomic Energy Authority Constabulary was replaced by the Civil Nuclear Constabulary. (Crown Copyright/National Archives)

# I

Rural England was not an unusual beat, but it was a quiet one. Having grown up in south-west England, I was familiar with the sights, sounds and smells of this part of the country. All, that is, except for a nuclear power station that was sitting on the banks of the UK's longest river. I had no idea until my deployment, in 2006, that such a place existed, just down the road from where I had spent most of my life. Built in 1967, this enormous edifice and its surrounding grounds had been providing power to the cities of Bristol and Bath, cooled by the tidal waters of the River Severn, for over forty years – since before I was born. As a police officer in the CNC (Civil Nuclear Constabulary), I had been assigned to provide round-the-clock armed protection to the facility, the nuclear fuel it held and the staff who operated its complex machines, which harnessed the power of the atom. In order to deter crime and prevent terrorism, the CNC also patrolled the surrounding roads and villages, bolstering the considerable physical defences of Oscar Yankee – as the power-generating site was informally known – with dedicated armed-response units.

Oldbury Nuclear Power Station is named after the small village of Oldbury-on-Severn, which, with its population of less than 800, is small even by British standards. In fact, more people worked in the power station than lived in the village; most commuted from the larger towns and cities in and around the Severn Valley. Far away from the metropolitan British cities of London, Manchester and Glasgow, this part of the country was characterised by rolling countryside, farms and tall hedgerows. It looked like the stereotypical UK that you might find on a social media post or a box of seasonal chocolates. Being out of the way, it was an ideal place to locate a nuclear power station. Its remoteness from almost every major population centre in the country meant there would at least be ample warning if something went wrong. The presence of such a large and complex facility, however, did not complement the rural feel of the nearby communities, which had been around since before the Domesday Book of 1086.

By the time I arrived at the OPU (Operational Policing Unit), based within the high metal fences and fortress-like security measures of the power station, there was a heavily armed police presence on duty at any given time. The Civil Nuclear Constabulary was less than a year old and trying to evolve from its predecessor, the UKAEAC (United Kingdom Atomic Energy Authority Constabulary), into a modern police force. Our police station was a Portakabin, our boss was a sergeant and the service as a whole was just beginning to expand – we had been playing game of catch-up ever since the threat from international terrorism had spiked in the wake of the invasions of Iraq and Afghanistan. In those early days, a lack of authorised firearms officers (AFOs), police officers who were qualified to patrol with firearms and respond to armed threats, was not just a CNC problem but an issue for the UK in general. Terrorist incidents in London and abroad had acted as a catalyst to strengthen the capabilities of the United Kingdom's specialist police officers, tasked with guarding nuclear facilities and nuclear fuel. As such, with an increase in funding, a new name and a healthy supply of new police officers, the CNC was deploying officers to the previously undefended nuclear power stations around England, Scotland and Wales – a task that was previously handled by the occasional visit from a volunteer police officer and unarmed security guards employed directly by the power companies.

Before the CNC came into existence, the UKAEAC was, by comparison to any other police service in the UK, very small. Many in the British public did not realise that alongside the county and city police forces of Great Britain there were and still are dedicated agencies with specific law-enforcement functions. The British Transport Police, for example, secures and patrols the railway networks of England, Scotland and Wales. The Ministry of Defence Police has responsibility over the vast military infrastructure of the UK, whereas nuclear policing and security is handled by the Civil Nuclear Constabulary. Prior to this, however, the UKAEAC was barely able to muster 1,000 officers and had a different and very specific role altogether.

Even within the policing community, few had heard of this unique constabulary. Those who had heard of them usually worked near a research site or fuel facility. Officers in Cumbria would disparagingly call their UKAEAC counterparts 'the factory police'. These constables only had authority on the sites to which they were deployed and could not patrol outside 'the wire'. They were essentially restricted to matters within the boundaries of their facilities. Initially armed with little more than revolvers and second-hand military submachine guns, these officers were a security force more than a police force. Without the need to patrol dedicated generating sites, they were deployed only to governmental nuclear projects and not private concerns such as at power stations, processing sites and universities. They existed to protect government property, not commercial property. Unbeknownst to many of them, however, the generation of nuclear power was a beneficial by-product of the core functions of these facilities. The Capenhurst facility in Cheshire, for example, was a uranium enrichment facility, whereas Chapelcross in Scotland was built to produce weapons-grade plutonium for the first generation of hydrogen bombs. The vast amount of energy generated during research was a useful, profitable but irrelevant outcome.

The atomic studies carried out in the UK during the Cold War era created an industry that few countries could rival. Money was being poured into the nuclear arms race on both sides of the Iron Curtain, and Britain was no exception. However, with memories of espionage and sabotage fresh in the minds of the scientists and politicians involved, the need to secure these facilities along with their cutting-edge research and dangerous materials became apparent as soon as the concepts for new nuclear projects went up on the drawing board. From the late 1950s until the late 1990s, the United Kingdom Atomic Energy Authority was therefore served by a police force dedicated to the protection of its own facilities. The regulators and officials that oversaw the industry could be satisfied that they had created a nuclear fiefdom free from hostile interference. This control even extended to universities and laboratories, where scientists were looking for the answers to questions nobody had yet asked. As for the power stations, these private businesses needed only a government licence to operate and full-time security team.

For many decades this was more than adequate. After all, the threat of nuclear war outshone the threat of nuclear terrorism. Authors and Hollywood movie producers had long since postulated the idea of nuclear material falling into the wrong hands, but the intelligence and capability assessments of terrorist groups at the time did not create a sense of urgency within the government. Even though the existence of terrorism perpetrated by the IRA was a significant part of the political climate during these decades, it was felt that private security teams and electrified fences were more than adequate for most facilities. This all changed with the global War on Terror, which saw an increase in international military interventions as well as a rise in worldwide extremism. Less than five years after the attacks on the Twin Towers in New York, it was apparent that the small and unique atomic constabulary would need to give way to a new and more sustainable entity that was better equipped to counter such complex threats. With greater legal powers, greater responsibility and greater funding, in 2005 the Civil Nuclear Constabulary took on the UKAEAC's responsibility for all of the United Kingdom's fuel storage, research and generation sites.

Funded by nuclear site operators, atomic research concerns and stakeholders within the nuclear industry, the Civil Nuclear Constabulary is perhaps the most unique police force in the British Isles. Attested and trained to the same standards as any other constabulary, officers from the CNC are specialised to respond to criminal and terrorist threats to the nuclear industry. Composed almost entirely of armed personnel, this police force is capable of deploying vast numbers of officers in response to any perceived risks to the many nuclear facilities and transport hubs within the UK. As a part of the UK's policing community, officers have also found themselves taking part in general duties, responding to active shooters and mitigating the ongoing threat from terrorism all over the UK and not just at nuclear facilities. The constabulary is also responsible for the security of nuclear material as it is moved around both the UK and the globe. This has seen British police officers deployed to France, the USA, Sweden and Japan as part of multinational security operations.

In the early days of an officer's career, it was common for the new officer or probationer to be paired up with a tutor constable – a more experienced officer who was expected to guide and mentor the 'probie' during their initial two years of service. My tutor, Dave, had a full year over my level of experience and was not even out of his own probationary period yet. However, being a capable and well-regarded officer, he had successfully attended and passed the tutor constable course. Outlining the patrol we were about to undertake, he assured me, 'We just need to get you familiar with the local area. These lanes and roads all look the bloody same so it's easy to get confused out there. There aren't any road signs either so I'm going to test your knowledge of the road names as we go around... It's no good calling for Avon and Somerset[1] when you need them but not being able to tell them where you are.'

Dave was in his late twenties and had previously been a club rep and lifeguard on the Spanish resort island of Ibiza. The ladies out there loved him and he knew it, but he had already settled down despite his young age. Blonde and charismatic, he had also been a talented rugby player before joining the police; his large and powerful build made his weapons seem more like toys than police equipment. He was a model police officer. Not 'model' as in ideal, but 'model' in that he could have come from a casting agency as opposed to the Constabulary Training Centre (CTC). He carried on talking as I got myself ready.

'So I reckon we'll be out there for maybe two hours. That's a bit longer than normal, for an external patrol, but we're going to stay near to the station anyway so that you get a good feel for the area.'

---

1 The CNC operates in the geographical areas of other constabularies. For example, Cumbria Police are responsible for the Sellafield area and Avon and Somerset Police are responsible for the Oldbury-on-Severn area.

Originally from the countryside surrounding the Black Country, Dave seemed like an interesting bloke. My guess was that he could have done anything he wanted with his life, but here he was providing armed security to a little-known and out-of-the-way nuclear site. I resolved to get to know him better, not just because he was my mentor but also because I liked his laid-back style of welcoming me into the team. First of all, though, I had to get to know my new beat.

Our patrol car, or ARV (armed response vehicle), was a large 4x4 sports utility vehicle, with police markings and lights. This made it virtually indistinguishable from the local forces' own police cars. As far as most people were concerned, while on patrol the CNC were a police service in the same way the local officers were. This meant that if we witnessed a crime or accident or needed to offer assistance, we should be able to at least provide a basic level of response or support.

'Nothing is going to happen out there,' noted Dave as he secured his G36C rifle in the vehicle gun box. 'Nothing ever does, especially on Saturday mornings. I'll drive this one, and when we go on patrol later, you can have a go, yeah?'

'Have you ever driven a police car before?' he asked teasingly.

He'd seen me arrive for my first shift in a dented old purple Ford Fiesta – a car certainly befitting my lowly status and young age. At twenty-two I had been the youngest recruit on my course and was now by far the youngest officer at the OPU. But I was also a veteran of the ongoing conflict in Iraq, having served there recently. As an infantryman I had taken part in vehicle patrols hundreds of times, although as a police officer this was my first time outside of training. For some reason, training at that time didn't cover the use of cars. Patrols at the training centre had always been on foot.

'Sounds good,' I replied. '... er, am I supposed to wear a seatbelt? It's just I can't get this seatbelt over all this stuff.'

I was wearing body armour, an equipment vest and had my own G36C resting across my chest as I made myself comfortable in the passenger seat and fumbled with the seatbelt. Dave looked at me for a few seconds before getting in and closing the door. Either he thought I was joking or was trying to figure out how this naive young officer had completed his training if he can't even manage getting in and out of a car by himself.

'Belt up and shut up,' he said with a wink. 'Let's go.'

It was an important day for me. My first shift in my new role.

Growing up in rural England I always knew I wanted to be either a soldier or police officer. As a boy, I used to enjoy visiting the local library and borrowing books about the military (the more pictures, the better). Both of my grandfathers had served in the military, and although they did not speak about it often, I admired and respected what they had done with their lives even though it wasn't through their own choices. On my father's side, my grandfather had been in the Royal Engineers and served during the North African Campaign as well as the invasion of Italy and southern Europe during the Second World War. His home was devoid of any hint of his military career; even his medals were lost in a box somewhere in the attic. There were no photographs and no mementos of that chapter of his life. Whenever my childish curiosity got the better of me and I asked about the war, he would say, 'The Germans were very brave,' and little else. On my mother's side, my grandfather had been stationed on the remote and inaccessible Christmas Island, in the South Pacific. This tiny speck of land barely covers 50 square miles and during the 1950s was used by the UK to test nuclear weapons in a project known as Operation Grapple. Far from prying Soviet eyes, the British were able to detonate four nuclear bombs in the space of a single year, refining and perfecting the world's first hydrogen bomb. He recounted the bomb tests, which not only regularly blew his accounting office over but also momentarily allowed him to view the bones in his own body as the nuclear flashes passed over him and the other personnel present during the detonations. He had gone on to have a full career in the Royal Air Force and learnt many skills that set him up for a second career after leaving the services.

My father, on the other hand, had not served in the military but was serving as a police officer. Joining the police in his mid-twenties, he had left a job at British Telecom in order to find better career prospects and a stable future on the thin blue line. His then wife, my mother, was pregnant with my older brother but he was still in need of a satisfying career – something that was not easy to come across in Gloucestershire at that time. Going on to become a uniformed police officer during the 1984/85 UK miners' strike, he had been injured in Yorkshire when a brick, thrown from a picket line, hit him on the head, knocking him unconscious and breaking the bone beneath his right eye. At that time the UK seemed to be going through a lot of changes. British politics was heavily polarised, and Margaret Thatcher's Conservative government was less than popular with the miners, workers and unions of the UK. With emerging technologies and complex societal issues, Britain was fighting to retain its status as a world player while also fighting to retain its industrial heritage.

With these social challenges, the Cold War and with the ongoing threat from the IRA combined, the Conservative government had started a police recruitment drive. Those recruited to respond to this new world order had no idea what was in store for them during their careers. Despite the overtime and arduous conditions faced by a copper in this period, in the summer of 1984 my father was able to increase the size of his clan with twins: myself and, thirty minutes later, my younger brother. As he eventually became a senior detective, we got used to his absence in the household and accepted it as a normal part of growing up. Indeed, he would go on to investigate one of the biggest murder cases in British history. By 1993, Fred and Rose West of 25 Cromwell Street, Gloucester had murdered at least thirteen people including members of their own family, in a case that shocked and intrigued the entire country. The details of the murders were horrific. Tortured and mutilated, the Wests' victims suffered unimaginable violence before they finally expired. The detectives – including my father – and the police officers of Gloucestershire Constabulary were tasked with investigating the murders, collating the evidence and seeing that Wests were brought to justice.

My father loved his work and I loved to hear about it. We would often enjoy evening walks together, rain or shine. He would recall anecdotes from his training, his days as a probationer, as a sergeant, as a detective and up through to a senior position in the constabulary. By now he headed up Gloucestershire's very own FBI, the MCU (Major Crimes Unit). At times it sounded as good to me as a life in the military, if not better. I was full of questions about what he was doing, what he had done recently and what he was planning to do next. Courses, investigations, multi-agency operations – anything I vaguely understood was up for questioning. Of course, he never mentioned at that time the drawbacks and sacrifices that a life in the police would sometimes bring and how it would affect his personal life. Gaining promotions and spending so much time away from his family eventually took its toll on the family. By the time my twin brother and I were in our early teens, he had decided to separate from my mother, leaving her at the family property with her three teenage sons.

As I was preparing for my final years of secondary school, the First Gulf War had recently ended and books about the conflict were becoming popular. By that time I was also delivering newspapers around the lanes and roads near Dillon's, the local newsagent. The British Army and their NATO allies were active in Kosovo, and I would keenly thumb through the tabloids before pushing them through letterboxes in order to glean the latest news from the frontlines. But despite my ongoing and adventurous ambitions to join up, there was a problem developing. It was my knee. During my mid-teens, while training to compete in the 200m sprint with my school's athletics team, I started experiencing sharp pains in my left leg. At the time everyone, including myself, thought that it was merely a strain that would heal itself given enough time. In the end, however, it took three years before I was able to run again.

Missing out on years of sports and fitness at such a young age meant that I shifted my focus to music and literature. I enjoyed studying foreign languages and writing about my accomplishments. Instead of playing cricket or rugby for my school, I began to study music and play guitar. Making new friends with interests in the performing arts diluted my desire to join the military or police for the time being, and I instead began to focus on other pursuits, not necessarily related to my previous ambitions. The school I was at had a huge music department with plenty of influence. Before I knew it I was playing Motown, country and the latest rock hits in shows around the county, and actively playing in groups outside of school. I thought it was great – I was a rock star in the making. The opportunity to take it further was only ever one show away, as far as I was concerned. At that time I could not have been further from a soldier or police officer if I had tried.

Moving into my late teens, however, raised the prospect of securing a regular job. The idea of further education at music college or a university didn't appeal to me. Many of my coworkers at the newsagent had been students, and although they talked a lot about what university life was like as well as their weekend activities, I couldn't detect any discernible difference between *working* all week and then going out at the weekend and *studying* all week and then going out. At least those who were working were spending money that they had earned and not borrowed in the form of a student loan, which more and more undergraduates seemed to be doing. But it was also obvious that they had enjoyed themselves during their studies. I knew that if I decided to choose that route, I would probably enjoy it as well. I was expecting good grades, certainly good enough for university. My school was even actively laying on trips to universities and assisting with the application processes of many students. Most of my friends were eagerly planning their next few years and trying to find out who had chosen what course – and, more importantly, where. My twin brother, too, had his eyes set in university. The then Prime Minister, Tony Blair, had famously referred to 'education, education, education' in his manifesto, and it seemed as though the message was getting through to many schools and students. There was no mention yet about conflict, conflict, conflict, which would be one of his more notable legacies.

'What about you, Matt?' said Robin, one of my closest friends at the time. 'Where do you fancy? Somewhere that does music?'

'Actually,' I replied, 'I think I might give the army a go.'

He looked puzzled and a bit amused. He knew full well that I had been studying music while the vast majority of students had been on the sports ground.

'You can't even run, Matty! Why don't you give uni a go? I'm looking at animation, Steve wants to do geography… Lucy's doing French… you'd probably ace that…'

He continued with all the information he had gathered from the common room, where the seventeen- and eighteen-year-old students relaxed while not studying or preparing university applications.

'No, I just don't feel like it's for me. And anyway, I can always go back to it if I feel like I made a mistake.'

He looked at me for a second longer and I knew exactly why. I didn't look like soldier material. I didn't carry myself well in public. I was tall and thin, with long hair and usually wearing dark, ill-fitted clothing. I looked like I belonged on a different planet, and nothing like a member of the military. Robin was right to look bemused. I hadn't run in years.

'I bet you a pint you won't go through with it!' he added, teasingly.

So if I was serious about the army then I would need to get fit, and soon. I would also need to get a haircut.

Nobody believed that I would be able to hack it in the army. Especially the infantry. I had been thinking about it constantly and even went to the recruiting office in the city to find out more. In the end I decided that I would join at the bottom and see what real soldiering was, and what life was like in the fighting arms of the ground forces. If I wanted to move up, move to a different branch or service, or leave entirely, I would cross that bridge when I came to it. Still keen on an evening stroll, I had called in to my father's place, a large townhouse he shared with his new wife and their black Labrador.

'You won't learn anything in the infantry,' my father told me.

Perhaps Tony Blair's mantra on education had got through to him as well. Or maybe he was thinking of what his own father experienced during the war. I disagreed. I knew that I probably would learn a lot; particularly about myself, what I was capable of and how to work harder than I have worked in my life. Maybe the recruitment literature had given me too idealistic a view of what I was setting myself up for. I was under no illusion that I was the typical infantry recruit. I also knew that I would learn about the world and about people. At least I hoped that I would. What he meant, I guessed, was that I wouldn't be able to make much money when I left because I couldn't repair technical equipment or fly a helicopter or manage groups of people. Pieces of paper seemed to mean a lot to him at that time, and a certificate saying 'combat infantryman' was 'not worth the paper it was printed on'. I had argued with him previously about my education choices and non-academic subjects, so I wasn't prepared to go through a second bout of paternal interference. Besides, he hadn't been home for the past four years. It was none of his business. I decided that was that. I had made up my mind.

'If my brother can decide to apply for thousands of pounds of debt to pursue higher education,' I reasoned, 'then I can decide to start my career where my interests lie.'

# 2

'You need to book us on with the locals, mate,' said Dave as we drove out of the power station to begin our patrol. 'Our lot already knows we're here, but you need to let Uniform X-ray know as well.'

Uniform X-ray was the callsign for the local area control room, who monitored police patrols and assigned tasks to available officers. I pushed the transmit button on my radio, waited a moment and sent my message.

'Uniform X-ray, from Oscar Yankee Zero One, over.'

They replied almost instantly. 'Oscar Yankee Zero One, go ahead.' The controller spoke quickly and efficiently and I could detect a slight south-western accent in her voice.

'Uniform X-ray, good morning, can you show Oscar Yankee Zero One as stat two, over?'

'All received.'

It was another first; my first radio call on my first patrol. But Dave looked less than impressed.

'It's not a bloody chatline, mate.'

I shot him a glance, unsure of what he meant.

'Just tell them we're on duty and that'll do. None of this "good morning, how are you, did you have a nice evening" rubbish. They've got better things to do than talk to you.'

By now we were picking up speed and leaving the power station behind us. He had a point, I suppose, but I didn't feel that I had overly eaten into the controller's time. There was barely any radio traffic and besides, what was wrong with showing some good manners? I shifted my attention back to the patrol. It was only the second time I had driven on this road, the first being on my way in earlier that morning. This time I was driving in the opposite direction, in an ARV and with enough firepower to deal with any armed incidents that might occur in the middle of rural England on a quiet Saturday morning.

A lot of a CNC officer's time is spent on patrol – both inside and outside the wire. On an external patrol, officers check around the immediate vicinity of their power station or nuclear site. The idea is to detect and disrupt any illegal or terrorist activities that could be linked to the area. For this reason, patrols don't usually stray more than 5 kilometres from the boundary of the facility. One of the other purposes of conducting externals was to create a relationship with the local community. In Oldbury's case, the local population of farmers and rural families was often more aware of what was going on in the area than we were. They could recognise an unfamiliar car or person instantly, and with good reason. Being so far out of the way, criminals who targeted these farms and houses knew that a police response would be slow in coming should the locals attempt to summon it. The community had always been self-sufficient, securing and policing their vast acres of farmland by themselves. But they appreciated the extra security that overt police patrols brought to the area, despite our role being primarily the safety and security of the power station and its fuel. It was no coincidence that, upon setting up our OPU, rural crime in the area dropped off almost completely in less than a year.

Internal patrols (patrols inside the wire) had much the same emphasis on overt security. Armed police officers are deployed to many critical sites in the UK to act as a deterrent and to respond instantly to any incidents that might occur. Anyone who has visited an airport, government property or royal household will no doubt have seen armed officers on duty, 365 days of the year. The nuclear industry is no different. However much of what goes on is never seen by the general public. Alongside the police patrols present at licensed nuclear sites, for instance, private security teams can also be found. Responsible for access control and matters related to the private company that owns the site, these security guards conduct their own patrols and ensure that the site licence conditions are met. These licence conditions are set by the ONR (Office for Nuclear Regulation), which employs inspectors to determine whether a site is permitted a licence and thus able to operate. This body was created through the same legislation that formed the CNC, establishing an independent nuclear regulator that falls outside of the larger and slow-moving Health and Safety Executive of the British Government. The ONR puts a great emphasis on the numbers of guards, cameras, fences and police officers that should be in place. The 'regulators', as we call them, also work with international partners, intelligence agencies and special forces to ensure the security of every asset in the UK's nuclear inventory. Most information concerning these matters is not in the public record; suffice to say, the defences of a nuclear power station or the precautions for the movement of fuel, for example, are substantial.

Patrolling the area felt new to me. I had a sense of insecurity and doubt, as if I was being assessed by my mentor and by the community. In a way I was; Dave was on patrol with me and showing me the ropes. These feelings were balanced by the excitement and satisfaction of getting to this point in my life. I was sure that, given time, I would relax into this new role and into policing. It reminded me of other 'firsts' that I had gone through, where I had felt the same way. The first day at school, the first day a new club or activity – or my first day in the army.

☆

Arriving for recruit training with the army on a Friday evening, I was certain I had made a mistake. In fact, I was certain that I had made more than one. Why did I join the army? And why did I join the infantry? My father was probably right. He wasn't being argumentative but probably anticipated I would not fit into this new kind of environment. I felt very out of place, like I had accidentally arrived at a party where I didn't know a single person. I didn't feel like I had a lot in common with any of my fellow recruits despite our shared ambitions and temporary uniforms. Here we all were, looking to get through training and to join our infantry regiments. The entire cadre was composed of young men going to units in south-west England: the Devon and Dorset Regiment, the Light Infantry, and in my case the Glosters.[1]

I was wearing a set of green coveralls that were probably two sizes too big, and a pair of white 'silver shadow' army trainers while standing awkwardly in a queue to have one of my first-ever meals away from home. The large mess and adjoining kitchen were in a wide, old-looking, single-storey building that looked like a relic from the Second World War. It probably was. The whitewashed brick walls and small, lead-lined windows didn't lend themselves to modern architecture. There was plenty of chatter as I waited my turn at the servery, but none of it was coming from my lips. Instead I was preoccupied with a hasty assessment of the food on offer. There was some sort of pinkish meat; probably pork or ham. There was another fleshy substance that was white; I guessed it might be fish. Shit. Where was the vegetarian option? Shit! What would people say when they found out I was vegetarian? Should I just try to eat some meat or maybe have a double helping of peas

1 The Royal Gloucestershire, Berkshire and Wiltshire Regiment (RGBW), a short-lived infantry regiment in the British Army, was often abbreviated to the word 'Glosters'.

in order to fill myself up? I hadn't even introduced myself yet, let alone been formally sworn into the military. While settling myself in at the ITC (Infantry Training Centre) I had not said a word to anyone. At eighteen years old and knowing little of the world, my insecurities were at that moment bouncing around my mind and down to the pit of my empty stomach. Too late.

'Is there a vegetarian option?' I squeaked to a less-than-interested cook standing behind the counter, spatula in hand. His chef whites had a small, single silver stripe on the breast, marking him out as a lance corporal. *I've definitely made a mistake*, I thought to myself. *If I was going to join the army then maybe I should have joined as a cook instead of the infantry. I would probably have been a good cook...*

Clearly this unanticipated question from an unimpressive recruit standing before a culinary NCO needed some thought before an answer could be given.

'No,' he finally sighed at me. So I just moved along the line, my tray disappointingly empty of any hot food.

*Peas and chips it is, then*, I thought, trying and failing to prevent myself from blushing and drawing any more attention to myself than was advisable this early in my career.

'Are you a veggie are you then?' Here it was, the first unofficial and somewhat social question I was asked, in my new role as a defender and faithful servant of the British people. I had just sat down opposite a pale-looking recruit with thinning brown hair who spoke while looking at my plate and not at me in particular.

'You can't be a veggie in the army,' he continued, while satisfying his healthy appetite with the pink meat I had seen earlier. Maybe he was right. I wasn't doing well so far, but I decided not to get into a debate about dietary preferences this early on. There would be plenty of time for that, probably sooner rather than later. I could already detect other recruits at the table pretending not to look at my bare meal made up of only two colours.

'Where are you going after this?' I asked, changing the subject to a more agreeable one that would hopefully involve a few more people than just me and my pale new friend. The question wasn't literal; I wanted to know which regiment he was aiming to join.

'Devons,' he said, mouth full, not looking up but now focussing on his own meal as opposed to mine. That seemed to do the trick. Other potential recruits for the Devon and Dorset Regiment began to chip in to this inaugural conversation. It seemed like the whole table was destined either for that regiment or its Cornish neighbour, the Light Infantry. *Great*, I thought to myself while listening to the rest of the table talk. *I'm the only Gloster here and the only vegetarian in the whole bloody army.*

At that time, basic training was split into two phases. Phase one was about soldiering skills in general, the things every soldier learns when they enter the military: how to march, shoot, read a map or render first aid, among other things. Phase two was for further training in the recruits' preferred branch of the army. In my case, this meant infantry skills. This phase of training was designed to provide the specialist knowledge the soldier needed before they could be assigned to their unit or regiment. As we were all destined for the combat arms of the military, rarely did we mix with any other parts of the army.

One day, Coffey, my pale friend with ambitions to join the Devon and Dorset Regiment (or D&Ds), was lying on his side, basking in the spring sun like a lizard under a lightbulb. We had come to HMS *Raleigh*, the Royal Navy's basic training establishment, to practise some shooting before going on exercise at Woodbury Common, a harsh and thorny heathland that regularly hosts training for the Royal Marines and Army Commandos. It was a warm spring morning, so being outside and not engaging in any strenuous activities felt quite pleasant. Only a few weeks into phase one of training, for many in our company this was one of their first times firing live ammunition. I had only previously used a shotgun, having visited the farms of family friends and taken part in some recreational shooting. Like many of the other members of our cadre, though, I had probably shot fewer than 100 poorly aimed rounds at this point. William Tell I was not. Yet here we were, practising our skills over increasing distances. However, supremely confident in his marksmanship and soldiering in general, Coffey was regaling our section of recruits with stories from his time in the Army Cadet Force. He didn't sound particularly like a grizzled and bitter veteran, but I got the impression that he thought he did.

'So yeah, they threw a dummy grenade into the top floor of the FIBUA house and after it popped, they ran in shooting.'

His latest anecdote was about the time he was in Kent, experiencing some urban combat or FIBUA (fighting in built-up areas) training.

'Afterwards I told them you can't throw a *real* grenade upstairs 'cos the blast would damage the floor and kill you as well.'

The section didn't particularly bother to acknowledge his latest assessment on military tactics and training; instead they focussed on their Cornish pasties, while I focussed on the meat-free equivalent. It was around eleven in the morning, and, being a thoroughly south-western cadre, ignoring elevenses[2] was considered as bad as skipping breakfast. Maybe worse. I wasn't sure what I was eating but ate it anyway.

'So really we won that one,' he concluded.

It all seemed ridiculous to me. None of us doubted that he had been an army cadet, as his drill and basic military knowledge were sound. I had been a cadet as well, but with the Air Training Corps and not the army. However, his talking about winners and losers as if combat was a simulation or game seemed to be immature at

2 Elevenses is a traditional mid-morning break in the UK, particularly enjoyed in the southern parts of England.

best. At that time the British Army was heavily engaged in Iraq and Afghanistan; the news was full of incidents where British soldiers had lost their lives. Barely a day went by without a news report stating that 'a British soldier has been killed'. Only days earlier, at Majar al-Kabir in Iraq, six Royal Military Police soldiers had been killed by an angry mob while cornered in their police station. I supposed that the Iraqis had 'won' that one, and the service personnel had lost. No doubt if Coffey was the umpire for that particular match, it would have gone differently. He could have called for a time-out or a slow-motion replay. Maybe he would have hit the pause button. Up to date with his latest war story, I swallowed my mid-morning slice of military-grade vegetarian option and made my way to the firing line to practise my marksmanship skills with the rest of the section.

Returning to the training centre a few days later, we had no time to shower or eat. 'Your weapon, your kit, yourself' was the mantra we had become used to. Cleaning our guns and equipment probably took longer than it should have with the training staff constantly reminding us that 'you're in your own time now', but as we were inexperienced, no matter how thoroughly we thought we had cleaned our guns, there was always a black mark, a smudge or rust somewhere to be found. Each time any of us approached one of the NCOs charged with supervising our development as soldiers, we failed to pass muster and subsequently missed out on a chance to have a leisurely shower and a few precious, stress-free minutes. No sooner had we finally satisfied the meticulous attention to detail that our overseers clearly possessed than we heard our next detail.

'Right!' Corporal Minchin instantly had our attention. 'Return your gats to the armoury, change into PT kit and get back down to the parade ground, with the speed of a thousand greased weasels!' I guessed a greased weasel was probably quite a fast mammal so didn't hang about.

It was a good idea to be first in the queue when returning weapons to the armoury. It was a laborious and time-consuming affair, with one gun at a time being handed to the armourer in his secure cage, where he checked the serial number and signed the issue/return register. Though our boogle of weasels moved fast, I doubted that the armourer would be as nimble. Being one of the first in line also meant more time to get changed and even a quick comfort break if you were lucky.

By the time I was back on the parade ground, resplendent in my thoroughly worn-in 'silver shadow' army issue trainers, I found that I was one of the first to return. Corporal Minchin promptly paired me up with a fellow recruit, Drewey, and told us to face off against each other, adding us to two ever-lengthening parallel lines. We knew what was coming. Press-ups and sit-ups, more than likely followed by a mile-and-a-half run to be completed in less than ten minutes and thirty seconds: the BPFA, or Basic Personal Fitness Assessment. I wasn't worried about this. Those several months prior to joining the army I had spent getting fit. One of the benefits of living in the Gloucestershire countryside was the abundance of space in which to train. I didn't need to go to a gym to build up my fitness. The abundant tracks and hills in and around my home were more than enough. I knew that I would be okay during this assessment, and that Drewey would be as well. A dark-haired and olive-skinned young man, he could have passed for a Sicilian or Corsican fisherman – that is, until he started talking and his broad Cornish accent betrayed his true heritage.

''ow many?' he asked surreptitiously.

I shook my head subtly to indicate that I didn't need him to artificially inflate my press-up count. It was not unheard of for recruits, when paired together for PT, to count the numbers of press-ups and sit-ups carried out during the course of the training. This number would then be dutifully reported to the PT instructor, who would note your results on the progress chart that he seemed to carry at all times. Of course, this system relied heavily on the honesty of the recruit to pass on an accurate record of the movements that were carried out.

'You?' I asked him.

'Nah,' he replied.

Looking further down the line, I was sure that similar conversations were happening. The instructors surely knew it too, despite our clandestine attempts at communicating with one another. They had probably seen it before and even done it themselves at some point. But so long as we put on a strong showing, with plenty of effort and little to no complaining, that always seemed to do the trick.

'Best effort then lads. You've got two minutes to do as many *proper* press-ups as you can.' He expected each recruit's chest to touch the curled-up fist placed underneath them by their training partner, who would count aloud the repetitions. Our knees were not allowed to touch the floor and we could only rest ourselves in the arms-extended position. A whistle blast.

'One, two, three,' counted Drewey as I began. 'Four, five, six…'

I was managing one press-up every two seconds and would have been more than happy if I could complete sixty before the time ran out.

'Fifty-six, fifty-seven, fifty-eight…'

Before I could complete the last two, the whistle blew again and we changed over for the next round. But fifty-eight was good enough for me. Next time I would crack the sixty barrier; I was sure of it. Completing the activity, the instructors took our scores. Alphabetically we announced the scores of our partners.

'Adams. Broslin. Coffey. Curzon. Drew.'

After each name, the counter delivered the score. We formed ourselves back into rank and file, waiting for the staff to dismiss us so that we could shower and prepare for the next activity, or hopefully take some time to sort ourselves out after our exercise on Woodbury Common.

'Coffey, go and see Mr Wood,' said the PTI. 'The rest of you, fuck off.'

The adjutant of the training wing, we rarely saw or spoke to Captain Wood except for non-training matters. He was a late-entry officer, meaning that he had been promoted to officer status only after having first served his time among the troops. Past the point of frontline deployment, he nonetheless knew the military system intricately, making him ideal for his role. He was a large man, and he always seemed to have a warm smile – a rare thing at ITC. He was more like a doctor with a good bedside manner than an army officer, and had no doubt experienced and resolved every conceivable issue a prospective soldier might encounter. It was Mr Wood who swore us all into the army and insisted that, should we encounter any problems while at ITC, his 'door was always open'. As the adjutant he was also in charge of personnel matters, so presumably Coffey had some paperwork to sort out or a query to answer.

Returning from our evening meal, Corporal Minchin called us all onto parade outside of our accommodation. There was nothing unusual in this other than the impromptu nature of it, but even that was not strange; we were getting used to the idea of the unexpected, and just went with the flow. Seeing as we had all returned to our uniforms and were ready for our next tasks, we quickly made our way outside and awaited to hear what was next.

'Coffey's been binned,' began the corporal. 'You lot wait out here whilst we empty his locker and gather up all his shit.'

He explained that for the past three weeks, Coffey had been providing bogus PT scores and had not once managed to achieve the speeds required for the mile-and-a-half run. In his embarrassment he had refused to return to his bunk and locker while the rest of the course was present, so the training staff were rapidly filling black bin liners with his possessions. Discussing it among ourselves after we had finished for the day, things became clear. We all knew that Coffey wasn't the fastest runner and had some work to do to achieve a more satisfactory level of fitness. However, it emerged that during PT sessions he was threatening his partner to ensure they recorded higher press-up and sit-up scores than he had actually achieved.

For the first few weeks, the threat of kit suddenly going missing and uniform being damaged was enough for this bully to get his own way. Maybe these had been his methods in life so far. However, one recruit had reported this to Captain Wood, and ever since Coffey had been under the eye of the instructors who, unbeknownst to him, were making an accurate assessment of his activities to date. Had he merely been unable to achieve the PT scores, no doubt he would have been given additional training or possibly, at the very worst, been sent to join a new course where he could continue to build up his strength and fitness. However, when challenged he had admitted to his methods and so he was subsequently asked to leave the course with the option to rejoin later. He had failed to meet not only the fitness standards for a career in uniform, but the moral standards as well. We weren't sad to see him go.

# 3

We had been patrolling for over an hour. Understandably it was very quiet. After all, it was early on a Saturday morning, and the local populace were likely still in their homes, deciding on what to do for the weekend. It was a beautiful day, with the perfect weather for a leisurely stroll or a spot of gardening. It had been raining overnight and the cool moisture of the farmlands created a layer of mist around our beat. Other than the occasional transmission on the radio and Dave's running commentary, this was a very straightforward patrol. The ARV cruised along comfortably.

My mind wandered as we began another lap of our patch. I had recently moved in with my girlfriend, who was studying to become a doctor, and I couldn't wait to get home to tell her about my first shift. Our new place was a little closer to the city of Gloucester than I was used to, so being back in the green and pleasant lands of Gloucestershire's countryside felt like being home. This was where I grew up, after all. Given that Dave had assured me that nothing would happen, I would have to try to find some interesting anecdotes to tell my girlfriend that night. She had already tolerated my stories from basic training, so was finally due an interesting pay-off.

The beat was split into two areas, divided by a main approach road that led to the power station. On the north were Shepperdine and Rockhampton; two farming communities that were connected both physically and culturally to the historic and royal town of Berkeley. To the south were the village of Oldbury-on-Severn and Thornbury, a larger town that just bordered our beat. These secluded conurbations were beautiful and unique in their own ways, although a visitor to the area would be forgiven for believing they were passing through the same village twice. Shepperdine, for example, had a stronger manufacturing output than the other villages, with a number of farms owning small workshops that made and repaired agricultural equipment. One of these farms even had a working M4 Sherman tank. Usually he rented it out for movies and TV shows, but he had agreed that in the event of a determined attack by the Panzergrenadiers or any other hostile unit we could borrow it in defence of our beat free of charge – providing he did the driving.

Separated only by local knowledge and no physical markers, the neighbouring village of Rockhampton was full of cattle that regularly impeded the flow of traffic. With a head of nearly 100 beasts crossing the road at most times of the day, there were only two options: turn around and go back the way you came, or chat with the farmer. More often than not we chose the latter option. We wanted to feel as if we were part of the community, and we wanted the community to know who we were and why we were there. Armed patrols are not a regular part of life in most parts of rural Britain; even places like South Armagh had seen their patrols reduced after the Good Friday Agreement.[1] The Royal Ulster Constabulary and their successors regularly carried firearms while on duty, much like the UKAEA and the CNC, but the overt deployment of firearms in the UK was

1 Signed in 1998 by the British Government and belligerent parties, the Good Friday Agreement sought to bring an end to the violence in Northern Ireland. The agreement saw the self-government of Northern Ireland and a subsequent reduction in armed patrols.

not and is not a common occurrence in British law enforcement, particularly outside of London.

The nearest town, Berkeley, used to have a main railway line but still maintained a unique train terminal that was used solely for the purpose of delivering and dispatching nuclear fuel to both our power station and also the dormant Berkeley power station, which stopped generating in the late 1980s. Close to Oldbury and using the same river for cooling operations, the site was in the process of being decommissioned – this is scheduled to take until 2079, meaning it will be undergoing decommissioning for longer than it generated power. The nearby terminal was often used to move nuclear waste and was technically known as a 'trans-shipment site' – a place where nuclear fuel was moved to and from. This tiny station had a powerful crane, a small platform and a hut with a kettle and microwave. The whole site was little more than half a football pitch in size. It was so well hidden that even the local inhabitants of the adjacent town did not know of its existence. As far as they were concerned, the last train to stop at Berkeley did so in 1964.

Policing of trans-shipment sites technically falls within the remit of the Civil Nuclear Constabulary, but at this particular location, with the movement of fuel being so infrequent, the ONR deemed that a fuel truck driver with a pickaxe handle was adequate security. This unknown rail terminal was less than a kilometre from the estates of Berkeley Castle, the feudal fortress that gave the town its identity. King Edward II, who met a painful end in 1327[2] while imprisoned there on the orders of his estranged wife, Isabella, is said to haunt the structure; supposedly his screams can

2 On 21 September 1327, Edward II was reportedly murdered by two men on the orders of his wife and her lover. The king was pinned face down on a bed, and 'a kind of horn or funnel was thrust into his fundament through which a red-hot spit was run up his bowels'. The assassination was an attempt to leave no visible marks of injury.

still be heard coming from inside the castle as his spirit endlessly relives its final moments. I guessed that more people had heard these disembodied screams than had seen nuclear material in transit.

Also in the south of our area was the castle town of Thornbury. This market town is dominated by a typical British high street lined with shops and services, with a supermarket at one end and a castle (now a high-class hotel), church and rectory at the other. The further you went down the high street the further back in time you seemed to go. Thornbury was just within our patrol area, and it was often necessary for us to visit in order to refuel the police cars, speak to the local coppers and maintain a visible presence on the few roads that led to and from the power station. Policing responsibility here typically fell to the host force, but we knew that anyone wanting to go to the power station would normally come through Thornbury, so along with our patrols in and around the facility, we would also keep an eye on the cars that were coming and going through the town, particularly at the start and end of the engineers' shifts.

When I started, however, it was still early on a Saturday and I could probably have counted the number of people I saw on one hand. By the time Dave had welcomed me, outlined the patrol and deployed us, the members of the skeleton crew that was operating over the weekend had already begun their shift. They had arrived at around the same time that I had, but in cars far more impressive than my well-used Ford Fiesta. Theirs probably started first time and had working electric windows, unlike my chariot. I found myself wondering how someone becomes a nuclear engineer. I doubted that it happened by accident; no Homer Simpsons here. For that matter, I hadn't become a nuclear police officer by accident either. It was a deliberate decision. Not feeling suitable for a full-time military career, I had initially applied to join Gloucestershire Constabulary. However, owing to recruitment pressures on the force and what was later proven to be a system of 'positive discrimination',[3] I had transferred my application to the CNC, who were recruiting new officers at the time.

Until the engineers finished and the night shift made its way in, there wasn't likely to be any significant traffic in our area – that is, if twenty cars going one direction and twenty more the other way counts as significant traffic. Dave was right when he said nothing happens on a Saturday morning. In fact the instructors at the Constabulary Training Centre (CTC) had made a point of reminding the potential new officers that routine work with the CNC is exactly that: routine.

'But once you understand the routine,' remarked one of the training team during my training course, 'you will instantly recognise when things aren't the way they should be.'

Now well into phase two training, life in the army was becoming enjoyable and a lot more comfortable. We were in Norfolk, on our final training exercise. I was glad to be this far south, as the hills and countryside in and around the Infantry Training Centre at Catterick were hard work at the best of times. In Thetford, there was barely a hill to be seen. The North Yorkshire training areas were all in use to such an extent that the possibility of running a Final Training Exercise (FTX) was out of the question, and the powers that be had moved the entire cadre to RAF Lakenheath in neighbouring Suffolk as we completed our training. Lucky us. Aside from the topographical benefits, one other advantage was

3 In 2006, Gloucestershire Constabulary admitted to the Commission for Racial Equality and Equal Opportunities that they had 'randomly deselected' 107 police applicants who had successfully passed the initial application process; all were white males.

that the airbase was in use by the US Air Force and as such had much better facilities on site than we were used to. There was a bowling alley, shops and even fast food far and above what the NAAFI were providing us. I tasted my first Budweiser and sampled my first Subway sandwich, courtesy of Uncle Sam. We had to take advantage of these amenities while we had the chance; after all, we weren't there for pleasure, we were there for the serious business of completing our training and finally becoming soldiers.

At this stage of our training, around 20 per cent of our number had been removed from the course for various reasons. Some failed to achieve certain competencies and were 'back squadded' to another course in order to have more training and another attempt at completing it. Among our own number, we had a few recruits from the cadres further up the chain as well, filling in for some of those we had lost through failure or fatigue. Some had been injured during training and were awaiting their chance to rejoin. Others had decided to leave, feeling that the military was not for them. I thought it was a brave decision and a sensible one at that. It was clear that they weren't always happy, and if the strain of army life was getting to them before they had even joined their regiments then it would be best to change direction sooner rather than later. Having spent months as part of a team, forming bonds and making new friends, it must have been a hard choice to make. If Coffey's episode had taught us about honesty and integrity, then these individuals were also teaching us something about personal courage – doing the right thing for yourself is not always easy.

Arriving home in Gloucestershire after training, I was glad to have some time off. Although I was starting to enjoy army life, being at home was still a welcome reprieve. Over the course of phases one and two, I had found out things about myself and what I was capable of; things that my friends at university were also learning alongside their undergraduate studies. As most of them were home for the winter break, we had gone to a pub that we all used to enjoy before embarking on our separate adventures. The Seven Springs, just outside of Cheltenham, was popular with younger clients like us; student loans and my meagre wages could happily stretch to a cheap pint and a hot meal. We were also familiar with some of the staff that worked there, many of them being drawn from the alumni of our school.

'What was it like then, Matty?' asked Robin.

'Dirty, cold and wet,' I replied, 'but I told you that I would do it and I believe that means you owe me a pint.'

He was more than happy to oblige. It was good to see my old friends again – Robin, Adam, Gareth. They all looked exactly as they had the year before. I supposed that I had changed the most, now with short hair, a high level of fitness and a prodigious appetite fuelled by months of physical activity. In truth, I knew that I had not done anything yet in the army – although I did not want to admit it to them. But neither did I want to talk about training as if I was a grizzled veteran, as I had heard people do on the firing ranges of HMS *Raleigh*. I was happy to hear about what they were doing and what university life was like. The assumptions I had made about university life while weighing up my options for leaving school seemed to have been accurate. It probably wasn't for me. I was happy that they were all enjoying their first year of further education; it sounded fun. They had an active social scene, plenty to see and lots of new friends. But I had that as well. I felt that if I had gone on to study, then in all likelihood I would still have harboured the same ambitions that I had just realised. With that, and after a final pint and some season's greetings, we went our separate ways, promising to meet up when we had all returned home once again.

# 4

Returning from patrol, the ARV had to pass through a huge airlock-style gate system that marked the external boundary of Oldbury power station. I took our security passes and presented them to the gate security staff, who logged us into their network as 'on site'. More than mere ID badges, these security passes also allowed the guards, from their control centre in the gatehouse, to identify who was on site, how long they had been there and to a certain extent where they were. Returning to the vehicle, we drove a few metres and passed the first gate. The siren and warning lights activated as the mechanism pulled the massive barrier clear of the entrance. I was sure it could probably stop a tank without even so much as a dent. Driving through, it began to shut behind us.

'So that's your first patrol done. We'll park the car up and seeing as we've got everything we need, we'll do an internal patrol next,' said Dave.

The gate behind the patrol car had by now completely closed, meaning that an identical one in front of us could now begin to slowly open. Nearby was our police station – or, more accurately, our police cabin. It took less than a minute to reach it from the site entrance. Dave parked the car near to the door and we disembarked, taking all of our equipment with us. As I readied myself for walking our beat, he completed the vehicle logbook, detailing the number of miles driven, the reason for the patrol and the areas we had visited.

'Ready? Don't forget to let Uniform X-ray know we're back on site.'

He motioned with his head our direction of travel and we began to patrol the roads that connected the various parts of the facility.

Our nuclear power station, like almost any other in the world, was dominated by two central reactor towers known as R1 and R2. These were connected to the turbines in the turbine hall – another massive building that housed the machinery that converted the nuclear fusion into usable power; the kind of power that keeps the lights on, not the other type of nuclear energy, which is rather more difficult to keep under control. All of this required huge amounts of cooling, so there was also a pier-type structure that scooped in vast amounts of water from the Severn Estuary in order to generate steam and therefore spin the turbines. In simple terms, the nuclear fuel boils the water from the river to create steam and this in turn spins the turbines, thus creating power. In reality, however, the whole operation is highly sophisticated. The sheer amount of machinery – cables, wires, buttons, consoles, big things, small things and complex things – is absolutely staggering. I couldn't believe that anyone would be able to understand it. But clearly someone did, as this station had been powering cities in south-west England for decades.

We headed west, towards the pier and the cooling facility.

'The place is usually busier during the week,' said my mentor. 'There's plenty of projects going on all of the time – repairs, maintenance, little experiments and things like that – so this is definitely a good time to get used to the layout of the site. We almost always patrol outside the buildings and the guard force takes care of inside.'

This much I already knew from training. Essentially private property, all the buildings and nuclear equipment belonged to the site owner and not the government. They employed their own people to look after their assets. I signalled my understanding as we made our way further into the power station. All sensitive areas of the site had strict access control, so the guard force was posted to ensure only those with the relevant clearance could get in. As hard as it was to get onto site without the relevant credentials, the closer you got to the reactors and fuel, the stricter the procedures became. Outside of these areas, we patrolled and maintained a visible armed presence in the same way that the UKAEAC had done previously.

Rounding the top of the site, we were met with four large cylinders, each the size of a passenger plane fuselage.

'Those are full of acid, mate,' said Dave. 'You don't want to be around when they're topping those up.'

The acid was used to clean the pipes and tracts that moved water and steam around the power station. Instantly able to remove rust and scaling from metal, the chemical concoction was also capable of causing considerable damage to unprotected skin or tissue. Beside each vat was an emergency shower, designed to dispense a large volume of water upon whoever activated it by simply standing on a footplate. I hoped nobody had ever needed to use one. On the other side of the road were a pair of steel chimneys. Each around 80 feet tall, they housed jet engines that gave the nuclear reactors adequate heat and energy to start the reaction sequence. The solid bunker behind them was full of aviation fuel, used to power the engines. Even without the nuclear material, this place had some serious hardware that we needed to look after.

After exploring the environs outside of the station and now the complexities of the power station itself, I was rapidly reviewing everything I had learned in the past few hours. This first patrol was as interesting as it was intimidating. Trying to focus on the core policing tasks while building a mental image of the site was giving me plenty to think about. But outside of my ongoing mental monologue, the power station was almost silent. The sounds of our footsteps were noticeable even above the buzz of the power cables that linked Oldbury to the UK's power grid.

We made our way down the central road that passed between the reactor building and administration block.

'Fancy a cuppa?' enquired Dave.

'You bet,' I replied. Patrolling while wearing body armour and uniform and toting police equipment and firearms is quite different to enjoying a pleasant stroll, and a quick brew seemed like just the ticket.

'In you go then.'

Dave held open the door to the admin building so that my equipment would not hit the glass panes, and followed me in. At the end of the corridor sat a vending machine that would soon have another regular customer. Dave reached into his pocket and put some money into the coin slot.

'Tea?' He had already ordered his drink and was waiting for it to finish pouring.

'Yeah, milk and two sugars, thanks.'

'Why did you join the CNC, then?' he asked, cup in hand. Blowing on his drink to cool it down a little, he took a sip as I began the short version of my journey to the thin blue line.

'I was in the army before this,' I began. 'I always fancied it, the army. But after getting back from Iraq, I didn't really feel like it was the job for me.'

He looked at me, no doubt expecting me to expand on my previous statement.

'Politics or policy aside, the problem for me was…' I took a sip of sweet tea hot tea to test the temperature and continued. 'Was that despite it being a decent job, there was a lot of bullshit that came with it. Work for work's sake or poorly conceived ideas that were pursued for no positive outcome. Nothing was up for debate and it was not a place where questions could be asked freely. I could tolerate it for a while but it's just not me. I thought that if I'm going to do a job, I want less of that and more control over what I'm doing.'

He smiled, spinning the tea in his cup a little. 'Plenty of bullshit here as well,' he said, half to me and half to himself. 'At least it won't be a surprise to you when it comes your way, I suppose.'

I had guessed as much. While at the CTC, the more experienced officers who were attending other courses had offered their unsolicited two cents at the police club bar, which we called 'the Moon Bar' owing to its lack of atmosphere. After training was finished for the day, over a well-earned drink, it was easy to either overhear or become caught up in conversations about how 'things used to be better', or 'things should be done this way instead'. The UKAEAC officers were now reviewing their policing skills and needed to incorporate new legislation and policy into their duties as custodians of the nation's most dangerous assets. The process of changing into the Civil Nuclear Constabulary was not just a matter of new stationery; there was a need for vast organisational change as well. None of their comments dampened my enthusiasm for joining, nor that of anyone else on my initial training course for that matter. We were the next generation of nuclear coppers.

'So which part of the army were you in?' Dave continued.

'Infantry. I was in the Glosters.'

He nodded, indicating that he had at least heard of this small formation that by now had by now been absorbed into The Rifles, a much larger regiment. 'You see much action?'

This was a standard question, having been asked dozens of times by friends and family alike. 'I saw enough' was usually the shortest and best answer to give to most queries of that nature. But my tea was still too hot to drink, and with a few minutes to kill, I decided to elaborate.

'There was this one time. We had to go on patrol, our section of eight guys,' I began. 'It was close to Basra Palace, closer to the lines than the external patrol that you and I have just done. All we had to do was patrol up from the rear gates to the bridge adjacent to the hospital and back. It was probably around 600 metres, so there and back was just over a kilometre. A simple, routine patrol. We had three guys from the IZP[1] with us; the fourth officer didn't turn up. We were simply there to support them while they looked for a white Toyota pickup truck that had a player in it – someone that they wanted to arrest. As far as they knew, the truck would be coming past the hospital at some point that evening.'

I could see that Dave understood what I was talking about. Our own compound at Oldbury was also a fortress, similar to Basra Palace. It was adjacent to a river, with long approach roads moving up to the car parks and main gate as well as security features to deter people from getting too close. I sensed he was visualising the same patrol I was talking about, but in the south of England and not the south of Iraq.

My tea was drinkable by now. I had never been able to drink scalding-hot tea like some people. So I took a sip and continued.

'The IZP were stopping every vehicle that vaguely fit the description of their player's motor. The traffic was heavy and backed up, so getting in and amongst the cars was no problem. Half of our patrol were down on the road with the cops and the other half had scaled an abandoned building to provide overwatch. Then all of a sudden all hell broke loose. One of the Iraqi police officers fell backwards from the roadside, and crawled two or three feet to the pavement after leaning through a car window. He was probably checking some ID or talking to the driver who then shot

1 IZP were the Iraqi federal police who had taken over internal security duties in the aftermath of the 2003 invasion of Iraq.

him. Our patrol commander who was next to him pointed at me and shouted… something.

'I understood that he wanted me to check this guy for injuries. I was the patrol medic and ran over to him, whilst the rest of the team pushed forward. The truck drove away over the pavement, the driver shooting indiscriminately as he went, but I don't think he hit anyone else. That stop had initiated one hell of a contact. I got down low and took out my medikit, which I carried on my back. No sooner had I done that than I heard the pavement disintegrating behind me. A shooter from a rooftop was either having a go at me or trying to finish off the casualty. Or both.'

Dave had finished his tea by now. He looked at me and put the cup into the bin but his gaze remained fixed to me as I sipped my tea and continued to answer his question about whether I had 'seen much action'.

I decided that now was a good time to wrap the story up. This was my first day on patrol, after all, and I was here to learn from him, not the other way around.

'He was hit in the shoulder. It was too far up the torso for a tourniquet so I tried to stop the bleeding with a field dressing and pressure from my hands. Luckily the hospital was nearby so I dragged him over to the guards at the gate and they took him inside.'

I took a final swig of my tea and we finished our break. 'The thing I learnt about a routine patrol is; it's routine for us but not for the people we're looking for.'

Stepping back out onto the road, after our brief interlude we made our way towards the back of the power station, adjacent to the River Severn. This was where all the machinery that kept Oldbury self-sufficient was located. Shipping containers, hangars, sheds and compounds all contained something useful. A relic of the Cold War, the plant was built with resilience in mind. Civil defence in the previous epoch had different priorities to that of the post-2000 era. There were also several backup power generators, a sewage treatment works and even a fresh water supply.

'If there's a zombie apocalypse, I'm coming straight here,' noted Dave as he continued to guide me around the beat. 'This area is well busy on a weekday, so be careful you don't get in the way. Engineers and staff are coming and going all the bloody time. If you get in their way then they complain to their union and then we get to hear about it from the sarge.'

I hadn't met the station sergeant yet. Aled, as he was known on site, was also a veteran, but from the Royal Air Force and not the army. He didn't come on patrol all that often and definitely not on weekends. His job was to set up the OPU and to work with the site management to integrate us into the facility. Site security manager Mr Surl, the executive in charge of the guard force, was apparently not happy at the arrival of CNC police officers at Oldbury. As far as he was concerned his security arrangements were more than satisfactory, and the idea of an independent security team over which he had no influence had caused more than a few meetings at a managerial level. Once it had been explained to Mr Surl that the police were there in a counter-terrorism capacity and not as security, he was a little more receptive to our arrival. He had instructed his guards to cooperate, but only as necessary. In addition, and of course in the interests of the safety of his team, they were instructed not to patrol with us on site in any capacity whatsoever.

Coming to the end of the patrol, we made our way past the reactor buildings and turbine hall. These huge structures dominated the entire site and surrounding landscape. The buildings were necessarily large in order to hold such huge machinery and to form a solid barrier between the nuclear processes and the outside environment. One of the departments of the power station whose members were constantly on duty was the health physics team. With their offices adjacent to the reactor building, much like us they carried out internal and external patrols, but they were looking for radiation and not crime or terrorism.

'New boy?' asked the duty physicist. He didn't look like a mad scientist or fit any of the stereotypes I had around what a nuclear engineer would look like. He was a middle-aged man with a shaved head, a beard and a substantial frame. He looked more like a Hell's Angel than a scientist, but I didn't recall seeing any Harley-Davidsons arrive at the power station that morning.

'Yeah, this is Matt. It's his first day and I'm giving him the grand tour. Any chance of going to look in R1?' Dave rested himself against the wall of the health physics office, to relieve the weight of his equipment temporarily.

'Not today, mate,' he replied. 'They're getting ready to shut her down and change the rods out next week. You working nights? I'll take you and the new lad in once it's all buttoned up.'

Dave looked at me and looked back to the engineer. If he wanted he could have insisted on entering the area, but that's not how we operated. We knew that the people working inside could do without a new police officer doing some sightseeing and we also knew that the area was totally secure with no need for us to go inside.

'Yeah, we're on Monday and Tuesday night.'

It was also not a regular part of the patrol routine, but it was important to be able to navigate the endless corridors and passages that formed the reactor buildings. R1 and R2 were identical. In the middle they were joined by a bunker-like structure that housed the fuel crane. This gargantuan device would move where it was needed in order to insert fuel into the machinery or remove it. Usually, if one of the reactors was in use, the other was undergoing maintenance; as far as I could gather, the engineers were getting ready to manage a switch over from R1 to R2 on Monday.

We arrived back at the police station at the same time as another pair of officers. JT and Honking John had been on a foot patrol outside of the site, checking the fence line and nature trails that surrounded the plant. Most nuclear sites sat in a nature reserve or similar green space owned by the site operators. It was supposed to reinforce the idea that nuclear energy was clean and not harmful to the surrounding area. Around Oldbury there was a long walking path, an orchard, two lakes and a birdwatching area that looked out over the river. JT and his partner had just got back from their checks and were kicking the mud off of their boots.

'That the probie?' asked John.

These two officers had joined the police before the CNC was created and had both worked at Sellafield, Europe's largest nuclear facility. They were used to the job by now and transferred down to Oldbury when the constabulary began to expand.

'Yeah, you were already out when we started. Had to get his pass and access sorted so we missed you during the daily brief with Harwell.'

JT wasn't a sergeant but was running the shift, being the most experienced of our team of four. As the acting supervisor, he remained in touch with our BCU (Basic Command Unit) at Harwell in Oxfordshire. From there the duty police inspector controlled the activities of the CNC in the south of the UK.

He looked at me.

'All going okay? I'm JT, this is John,' he said, indicating to the officer who was just settling down at the table with a flask of coffee. Other than their heights, it would have been hard to tell the two apart. Honking John looked like a shorter version of JT, goatee and all. If someone had told me they were brothers, I would have believed them. John got his nickname from his need to 'honk' or complain about everything. From the weather to the quality of his lunch, Honking John had a lot of opinions – mostly negative. The only thing he didn't honk about was his role as a drunk in a *Lord of the Rings* movie. It was also a play on his actual surname.

'Yeah, it's going pretty well, thanks. I'm Matt,' I replied, wondering whether to sit down without my new colleagues asking. I decided that I would busy myself by removing my armour and getting the weight of my patrol equipment off before joining them around the table.

'Okay, then,' said JT as he sat down at the office table. 'Aled hasn't left us anything out of the ordinary to take care of over the weekend, and there's nothing on site that requires our attention.'

I was waiting for a 'but'.

'... So me and John will go on internal patrol now and you two can have some lunch. When we get back, you can go out on externals again.'

We were all sat around the single table on four plastic chairs that formed the meeting, resting, administrative and eating area of our cabin. There was a waist-high fridge with a television on top of it, adjacent to the head of the table; a kettle on a stool and a phone charger hanging limply from a wall socket. The other end of the cabin had a sink and a drainage board. Between the two ends of our Portakabin were four weapon-storage safes and two long rows of lockers; one for each officer who was assigned to Oldbury. Fitting four coppers into this tiny office was not exactly comfortable or convenient, but it was all we had at that time. Mr Surl had decided that as 'contactors' we should not have an area inside the administration building, despite numerous vacant offices. Instead, just like the cleaners and scaffolders who regularly visited the power station, we were not to be seen as receiving preferential treatment in the use of the site's facilities.

I'd been stationed in much less comfortable surroundings, so as far as I was concerned this was all great. I was finally a police officer, and I had finished my first patrol.

# 5

I had been at Oldbury for around three months and was starting to settle in. By now I had met the other officers at the station and had worked with most of them. They were a mixed bunch. Some were on secondment from other units, but most were based there permanently. We also received regular visits from the TRG (Tactical Response Group). These police officers were based in Oxfordshire but were deployed all over the south of England on a rotating basis. This meant that they had working knowledge of all the power stations in the BCU: Sizewell, Dungeness and Oldbury, as well as the Harwell research site. The idea was that if extra officers were required, those who turned up wouldn't need to be given any local training; they could just fit into where they were needed.

Returning from an external patrol with Cal, we saw that the TRG patrol vehicle was parked next to the police cabin. Upon entering we saw one of their officers doing a handstand against a bank of lockers. Before we could ask what was going on, Jodie from the TRG explained, 'Pete's had the hiccups for about four hours straight so he thought standing upside down might help.' I looked at Pete while getting my Guns N' Roses coffee mug out of my own locker, which was about a metre to the left of him. Pete was in his forties and a keen angler. Spending a lot of his time outside and constantly smoking roll-up cigarettes he looked older than his years. He was a former UKAEAC bobby and knew his stuff. But he didn't say anything. He just looked up from his position against the lockers, his eyes slightly red from what presumably was quite a prolonged effort at being upside down.

'Is it working?' I asked.

As if to answer before anyone else could offer an assessment on this treatment, there was a croak from Pete, who, in his inverted position, rolled his eyes towards the floor. His feet swooped down and his heavy police boots banged against the cabin floor.

'Fuckin' ridiculous,' he said to nobody in particular. 'Fresh air is what I need.'

He probably meant a smoke. Smoking on site was strictly prohibited given all the explosive gases and dangerous chemicals, so he disguised his desire for a quick fix by asking if anyone wanted to join him on an external foot patrol. I was filling up my mug with a quick brew.

'Yeah,' I replied. 'Let me have a quick cuppa and I'll come with you.'

We left the power station through a series of full-body turnstiles adjacent to the gatehouse. The guards inside could hear and see everyone that came and went so I stopped at their window to have a quick chat.

'Just off around the lagoons. Should only take ninety minutes or so.'

The guard waved his acknowledgment from behind a desk that supported a dozen CCTV consoles and got back to his crossword puzzle. The power station had two silt lagoons next to it. These enormous bodies of water were designed to receive any mud or silt that was accidentally scooped into the cooling system through the pier. An unintended consequence of this nutrient-rich byproduct and constant water supply was that patrolling the lagoons was like patrolling in the jungle, with waist-high vegetation, tall trees, the constant sound of insects and a microclimate that was suffocatingly hot in the summer. The ground was also very soft, which meant that patrolling in a vehicle was rarely possible. Given the proximity of this area to the power station and the overwatch positions of the roads around Oldbury it afforded, our OPU checked the lagoons regularly.

The north lagoon was the larger of the two. About the size of a large sports stadium, stands and all. The entire circumference was close to one and a half kilometres. Apparently it had become home to a rare kind of newt, which meant that from time to time biologists would turn up in their waders and stroll into the water in order to research the little salamanders. Normally this wasn't a problem but occasionally, when conducting research in their olive-green outdoor kit, with tripod-mounted cameras pointing in the direction of the site and radios to talk to their mates on the other side of the lagoon, CNC officers would have to deploy from the power station to find out what was going on. However, normally they would let us know if they were visiting and how long they would take. The area seemed quiet as we made our way onto the track that linked the power station to the lagoon. We headed into the vegetation and began our patrol.

Once sure he was out of sight, Pete reached into his belt pouch and produced a tobacco tin full of pre-rolled cigarettes. He was still hiccupping so was trying not to speak too much; he didn't want to annoy himself any further with his involuntary contractions interrupting his conversation. Instead, he sparked up his cigarette and patrolled down the track, a few steps behind me. We were wearing our full kit for conducting an armed response: body armour, G36C, Glock 17 pistol, spare ammunition, baton, handcuffs and pocketbooks. The only things we weren't carrying were helmets. It reminded me of a military patrol and was nothing like the training scenarios I had undertaken at the CTC in my shirt, tie, hi-vis jacket and bobby hat.

We weren't likely to bump into anyone here, but that didn't mean it never happened. Well out of the way and with a large staff car park nearby, the track was sometimes visited by adventurous couples who would arrive on weekends or evenings to explore the area – or each other. The Berkeley Hunt also rode through on their horses, following a scent trail. On one occasion I saw a white Ford Transit van trundling up the track while on patrol with Nix, a female sergeant who was on secondment to our unit. Catching up with the van on foot, we found that the driver and his mate had stolen a hefty amount of red diesel[1] from their boss, one of the local farmers whose property backed onto the north lagoon. To avoid the security cameras at the farm gate, they had entered and subsequently tried to leave via the lagoon. It was not their lucky day.

Patrolling with Pete, I heard a zip and the sound of liquid pouring onto foliage. 'Pete must be having a quick comfort break,' I thought. We were in the middle of nowhere and hadn't seen anyone, so it didn't come across as unreasonable. He was only a few steps behind me, and his proximity was further betrayed by his hiccups. Strangely, though, as I continued to walk, the sound didn't get further away. 'That's odd,' I thought as I turned, half expecting to see a water bottle being emptied or some other rational explanation for this localised downpour. There was Pete, cigarette in mouth, hands on his weapon and moving along in step with my slow pace. As it turns out, Pete had the strange skill of being able to patrol and piss at the same time.

'Do you mind?!' he said, slightly indignant. 'I'm havin' a piss for fuck's sake. Keep your eyes to the front.'

1 Red diesel is normal diesel fuel that is coloured cherry red to distinguish it from regular fuel. It is low-tax and permitted for use in off-road and agricultural vehicles.

Dutifully I resumed patrol. Somehow he had managed to avoid splashing his boots and trousers. I didn't ask him about his technique. Apparently now free from bodily discomfort and hiccups, Pete readjusted himself and we moved up onto the riverbank at the top of the lagoon.

The north lagoon was joined to the southern lagoon by a track that bisects that site and makes up part of the Severn Way, a 224-mile footpath that is popular with walkers and runners. The path actually cuts between the cooling water intakes and the main structures of the power station, necessitating the building of a subway beneath to keep the two areas connected. Members of the public often walked past, and we got to know many of the regulars who strolled by with their dogs two or three times a day.

As we passed between the pier and the power station, we could see the huge water scoops of the cooling system dipping into the river and pulling out tons of water every few seconds. The scoops were each the size of a small car, and rotating on a vertical conveyor belt that never seemed to stop. As the water reached the top of the chain, the scoop inverted before making its way back towards the water, dumping its contents into the water intake, which was pumped over to the reactors. Any silt, weeds or unwanted contents were separated and pumped into the south lagoon, the next stop on our patrol. Smaller than the northern lagoon, these southern marshlands were easier to patrol. The high banks around the perimeter were firm underfoot and a lot more spacious. The site maintenance team regularly cut away any excess vegetation as the health physics team regularly patrolled the area, checking for any signs of unexpected radiation. Coming down off the banks, we rejoined the main approach road to the power station. We had to cross a small canal built to expel wastewater from the site into the public sewage system.

About two weeks before, I had been carrying out a similar patrol with another new officer, Mark, and looked into the canal just as a matter of curiosity.

'Mark?' I asked slowly. 'Why is the canal water bright green?'

He caught up with me and looked in.

'Fuckin' hell, Matt,' he said with his broad Cockney accent. 'That don't look right. Get onto health physics and get 'em down here toot sweet. I'll have a look around and see if some joker has dumped some dye or something in there.'

It was a good idea to check. Anti-nuclear activists had previously used projectors and lasers to display protest symbols on the cooling towers and the local protest group often voiced their objections at site stakeholder meetings. The water was a vivid and unnatural green. Using a short-range VHF radio that we always carried on patrol, I contacted the gatehouse to make the guards aware of the situation and asked them to bring out one of the physicists.

'Nobody else here, Matt, and no tins or packets. That green shit is probably coming from the site.'

Mark checked his watch and made a note in his pocketbook about this unusual occurrence, then let the duty inspector at Harwell know that our patrol would not be returning as expected but had gone static at the canal, awaiting the team from site.

A security vehicle arrived with three guards, along with a truck that the health physics team used to carry out their patrols. Mark had gone a little closer to the bank of the canal and was leaning over to have a good look in. 'That's making my face sting a bit,' he said. I hadn't felt any discomfort, but then again I hadn't been that close to the water. My first reaction upon seeing the green liquid was to keep a reasonable distance. It had been drilled into us at training that shielding and distance were the most effective defences against radiation, and if that stuff was leaping[2] I wanted to be nowhere near it.

'Well note it down then, so if something happens you can tell the medics when it happened,' I advised.

Aled, the station sergeant, had come out too and had a quick look into the canal. The guards had by now shut the vehicle barrier across the car park, and Aled put Mark and me a little further up the road to stop any oncoming traffic. The health physics team needed some space to work and to get themselves down to the level of the water.

'That water caused my face to sting a bit, Aled,' said Mark.

Aled looked at him, and then to me.

'No, I feel fine.' I was getting out my own notebook to take down a few notes as well. I checked my watch and noted the time, guessing it had been only ten minutes since we had looked into the canal. The site staff had been quite rapid in getting to us.

Soon after, we heard the vehicle barrier scraping open and Aled went over to talk to the site team. They weren't close enough to hear but I could see his shoulders moving up and down as if he was laughing. Aled was a big man with broad shoulders and looked like a stereotypical policeman. He had a greying goatee and shaved head, and a penetrating gaze that made him look fierce. This towering Welshman was in actual fact a very personable and likeable bloke. Appearances can be deceiving. He also got on well with the site staff, and knew all of the guards by name. Judging by his body language, this wasn't some sort of nuclear disaster. He came back.

'Yesterday,' he began in his soft Welsh accent, 'some of the engineers detected a leak in the plumbing. They couldn't figure

2 'Leaping' is a colloquial term for radioactive. The presence of radiation causes the Geiger counter or similar radiation detector to 'leap'.

out where the leak was so they put in a dye to find the rupture. When the dye didn't come out they figured that they must have got it wrong.'

That was a relief.

'How come the canal's bright green, then?' enquired Mark.

'The dye made its way through the system and eventually ended up in the canal,' he replied. 'Health physics just took a sample and detected nothing unusual except the colour. No idea why your face is stinging though. That bloke's in the water now, up to his balls in it, and he feels fine. You can go to the doc on site if you like though. I'll cover for you whilst you disarm and get yourself checked out.'

The site had a comprehensive medical facility that could deal with just about any incident that came its way. Dr Ludlow, the site medical officer, was also the father of one of my schoolfriends, who had gone to university in nearby Bristol. I knew that he wouldn't judge Mark if he felt the need to have a quick once over.

'Nah, I feel fine now,' said Mark. 'Let's get back to it, right Matt?'

After crossing the canal – now full of clear, fresh water – Pete and I finished our patrol and sat down in the station for refs.[3] He used his radio to let Harwell know we were 'stat four' – taking refreshments. He turned on the TV and adjusted the dinner fork that we used to improve the signal.

'Love a bit of Bargain Hunt,' he said.

I could never get into daytime television shows so got out a book instead. Cal and Fish had just come in from patrol as well and

3 Refs, or refreshments is the time allocated during a shift where the officer can take a break or undertake personal administration.

were taking a break, meaning that the other half of the TRG were still out. Calvin was a Gloucester boy through and through. Other than a brief spell in San Francisco when his wife was working there, Cal, much like me, had spent most of his life living and working in rural Gloucestershire. He was just about to turn forty, and was planning to take his family to Disney World in Florida to celebrate. It was always fun to work with Cal as he always seemed to be in a good mood. Even his bad moods were good moods.

I first met Cal at the CTC when he was on a course and I was a brand-new recruit. While in the Moon Bar he noticed that I was wearing a local football top and came over, pint in hand, to introduce himself to me and Mark.

'You goin' to Ol'bury?' he asked.

I hadn't even answered when he began commenting on my shirt.

'That's a Forest Green top innit. You a Stroudy?'

He was referring to the town of Stroud, near to where Forest Green Rovers played football.

'Yeah. I lived there most of my life,' I replied.

'You never 'av. I'm from there as well! Which part is home?'

His local dialect was completely understandable to me but I felt maybe Mark, who came from London, was struggling to decipher the strong accent.

'Stonehouse – well, King's Stanley, which is near there.'

He grinned and stuck his hand out. 'I'm from Leonard Stanley,' he beamed. Leonard Stanley was the next village over to where I was living. 'We're neighbours! Cal's the name,' he said, 'let me introduce you to some of the rest of the crew.'

Three Oldbury officers had come to the CTC while we were still completing our basic training. They were on a course at the firearms training facility to learn how to deploy the baton gun – a firearm that discharges a rubber bullet and is designed to knock someone down as opposed to killing or seriously injuring them. At the time, only power station officers were trained in its use. For that reason is was not part of the firearms syllabus for new CNC officers.

'This is Matt and Mark,' he said as we joined his group at their table. 'They're coming to Oscar Yankee once they finished up here.' Sat with a packet of crisps and a pint each were Fish and Gav.

Fish was a tall bloke in his mid-twenties. He had short, curly black hair and didn't seem to speak too much. He was one of the senior constables at Oldbury, having joined the while the constabulary was still called the UKAEAC. Gavin, an RAF veteran, was a cheerful guy in his thirties. He didn't look as athletic as the other officers at the table and was probably starting to lose the inch war that apparently begins after someone turns thirty. He nodded and smiled, raised his pint then asked us when we were expecting to complete our training. He was the acting sergeant of B shift – filling in as one of Oldbury's supervisors while the CNC tried to recruit more officers and promote suitably qualified PCs through the ranks.

'We have four shifts at site,' he explained. 'A, B, C and D. Nix is the A shift boss, I'm running B shift. A constable called JT is running C and D is run by Aled, who is also the station sergeant.'

It was great to hear all this. It was real experience from the ground, something that we lacked. Despite the worries that we had all felt during training, it was nice to sit and speak to the people we were hoping to work with in the near future. Training at the CTC focussed more on the policing of the major nuclear sites such as Harwell, Dounreay and Sellafield. On our intake, only five out of thirty officers were slated to be deployed at power stations.

'When did you start? Are you far in?'

Mark had returned to the bar to refill his pint so I answered for the pair of us.

'This is only week three, so we have twelve more to do.' At that time the IFC, or Initial Firearms Course, was fourteen weeks long. There was an additional week for those being deployed to power stations.

'Going alright, is it?' asked Fish.

'I guess so. We're going to do some stop-and-search training this week and our PST (personal safety training) as well.'

Cal, one of the more recent additions to the unit, was reassuring. 'Don't you worry 'bout that. It'll fly by. Piece of piss it is.'

# 6

Training for the CNC at that time was carried out at Summergrove Hall in Cumbria. This estate had once been an impressive Georgian Mansion, but after changing ownership several times and falling into disrepair, the site was bought by BNFL (British Nuclear Fuels Limited) and demolished in the 1980s. In its place they built a training and conferencing venue complete with canteen, accommodation and parking only a few miles from their main base of operations, Sellafield. Most of the site was used by the various engineers and scientists who were visiting Europe's largest nuclear facility, but B wing, or the Buttermere wing, was used as the CTC. The four main buildings were all named after famous bodies of water in the Lake District: Buttermere, Coniston and Derwent. As no lake begins with the letter 'A', Ullswater was the fourth block. U, B, C and D.

I arrived at Summergrove on a Sunday afternoon in 2006. My first day of policing was to begin the following day, which was 11 September. It was a warm Sunday, and the drive from Gloucestershire to Cumbria had taken me almost seven hours including a few breaks. Memories of recruit training in the army were at the forefront of my mind, so I had come prepared with all the useful pieces of kit that helped to make life easier in the military: travel iron, sewing kit, boot polish and brush – none of this was mentioned in the pre-joining notes, but I didn't want to have to make any unscheduled trips into Whitehaven, which hosted the nearest supermarket. I also brought some custom-fitting earplugs. Anticipating barracks-style accommodation or shared rooms at best, I knew how difficult I would find it to sleep with even the slightest noise from those sharing the billet. I had always been a light sleeper. Suitcase in hand, I shut the boot of my car and followed the signs to reception from the walled car park that was oddly devoid of any police vehicles, police officers or anything police-like in general. I thought I might be in the wrong place. Summergrove didn't look like a training wing or a police college; it certainly didn't feel like a centre for excellence in policing and uniformed service. It looked and felt like a budget hotel.

'I'm here for the CNC course,' I announced to the receptionist, an attractive, slim lady with had long, dark hair and dark eyes. Luckily for me she answered with a polite smile that disarmed any doubts I had about being in the wrong place.

'What's your name, love?'

There wasn't a computer system to check me in, but she efficiently flicked through some cards to find my name.

'Coniston, room five,' she said, handing me a key and a slip of paper with the passcode for a security door. The C wing of Summergrove would be the recruit accommodation for the entire course, but I was yet to meet any of my fellow officers. 'Head up to the bar at four o'clock, the training team will meet you there for orientation. Just follow the stairs behind reception and go through the double doors at the end of the hallway.'

*This can't be right*, I thought as I entered my room. I wasn't sharing with anyone. I had my own shower. I even had my own desk. I looked at the paper slip with my name on it and checked the number on the door again. Maybe the pretty receptionist had mistaken me for a technical visitor and not a police recruit. While it wasn't exactly the pinnacle of luxury, the room was a lot better than I had been expecting. I unpacked and squared away my kit and clothes. I had a picture of my girlfriend, which I stuck to the mirror above the desk, along with some sealed letters she had given me, to be opened in strict order on the dates indicated on the envelopes. I opened the window for a look around outside. Gravel tracks, trees and fields. *I've definitely had worse*, I decided. The quiet surroundings reminded me of home, reinforcing my feeling that Summergrove was an odd setup for a police training facility. But my less-than-comprehensive twenty-minute assessment of the place was by no means accurate. Closing the window, I decided to leave the room early and look for the bar, where I hoped to meet at least one or two fellow recruits.

I went up the staircase adjacent to the reception area. The pretty receptionist was still on duty, but apparently busy with some paperwork. Heading up to the second floor, there were a couple of payphones on the wall and a short corridor that headed towards a function room and the bar. Seeing a few people were already in the function room, I decided to go in and join them. Still no uniforms; no signs of law and order. I hoped that I hadn't dropped in unexpectedly at someone's birthday party. Probably not; it was too quiet and subdued. Maybe I had found myself at a wake. There were a few quiet conversations going on, as well as the odd nervous glance towards the large double doors I had just entered. I had planned to sit quietly on my own, but instead I sat down at a table where two blokes were speaking.

'Are you here for the CNC?' I asked, before I had even completed manoeuvring myself into a comfortable position.

One of them nodded but didn't say anything. He kept nodding for quite a while after my query. He was a big man and seemed noticeably awkward. As he sat, I could see that his feet pointed inwards with the toes angled towards each other. He looked like a blonde Lurch, the butler from the Addams Family.

'Aye, I'm Ewan,' said the other. Ewan seemed to be a little older than most others in the room and I guessed he was already in his forties. He had a thick Scottish accent and offered his hand.

'Robert,' said the other voice.

Introductions completed, albeit awkwardly, we proceeded with the inoffensive line of questioning that inevitably comes with meeting a new colleague for the first time.

It wasn't long before the room was full, well before four o'clock, with thirty new recruits checking their watches regularly and looking around, sizing up the others on the course. Just before the appointed hour, a troop of police officers came in: white shirts, black ties, numbered epaulettes and hats under arms. *This is it, then*, I thought.

Up until this point, joining the police had been a matter of paperwork and an assessment weekend. A few months previously I had gone to Ryton, near Coventry, to attend some interviews and scenario-based testing that determined an applicant's suitability for joining the police. This training school hosted many recruits from different constabularies under a central training authority. It also ran courses on behalf of other agencies, such as the CNC and MDP (Ministry of Defence Police). Following a battery of assessments and evaluations, there was a basic fitness test run by a short and angry PTI who clearly didn't want to be there. I remembered that he was wearing a UKAEAC T-shirt and a pair of Ronhill running leggings. He looked like he had been stuck in the early 90s for quite some time. His apparent frustration was highlighted by his treatment of the candidates. Blunt and borderline offensive, he wasn't a great ambassador for the service.

The officers who had arrived at the CTC function room looked altogether more professional and motivated. Without saying a word, they had our attention. Bringing up the front was an inspector who had a sergeant and three constables with her. With the new recruits sat around in various groups, we quickly assembled ourselves into one larger group, anticipating our initial briefing as police officers.

The sergeant told us to stand up. Dark-haired but greying, he was a solidly built man with a moustache that seemed to add to his authority. It was obvious that he had been in this line of work for quite some time; he carried himself with the kind of confidence that only comes after years of experience, having encountered most if not all issues that the UKAEAC or CTC could generate.

'I'm Sergeant Buckden, the training sergeant,' he announced. 'These are your instructors. You will refer to them as "staff" and you will call me "sergeant".'

We sat down and the staff took turns to briefly introduce themselves. Then the sergeant spoke again.

'You'll meet the rest of the training team tomorrow,' he said. 'These are your lead instructors. Go to them if you have any issues.' The instructors had all been in the police for years and were veterans of the UKAEAC. Except for the comfortable surroundings, this kind of orientation I had seen before, in the army. Buckden turned to the inspector and made way at the front of the room as if he had only been the warm-up act.

'Ma'am,' he said deferentially as he stepped back to join the other training staff.

The inspector thanked the sergeant before directing her attention to the thirty recruits gathered before her.

'Now that you have met my team, allow me to introduce myself. I'm Inspector Preston.'

She was a slim lady in her fifties with short, grey hair. She wore a police skirt and shoes while the rest of the team wore trousers and boots. I guessed that maybe, as an inspector, there were different uniform standards, similar to the distinction between NCOs and officers in the military. Sounding almost bored, she began to speak again.

'All of the day-to-day running of this course, Civil Nuclear Constabulary Course Number Five, is handled by the training staff you have just met. The rest of my team you will meet in due course. As stated by Sergeant Buckden, if you have any problems then you should see them in the first instance. If for any reason this is not possible, then you can come and see me.'

I didn't like the idea of a one-to-one with her. She seemed like an overworked schoolmistress or stressed-out librarian.

She then paused to look at the entire audience.

'Tomorrow' – another pause – 'you will begin your training, but the assessment of your behaviour and suitability to become an armed police officer has already begun. Everyone here knows who you are. Everyone locally knows who you are. I know who you are. There is very little you can do that will escape our notice.'

This subtle warning was not included by accident. I wondered what kinds of trouble the previous courses had gotten into. Surely those who were trying to become police officers would be on their best behaviour at all times, I thought. But then I recalled *The Secret Policeman*, a documentary from a few years previously, where an undercover reporter had infiltrated the Greater Manchester Police (GMP) as a recruit. His revelations about racism saw many officers dismissed and even the GMP recruit-training team reassigned.

'You will have the rest of the evening to settle in,' she continued. 'The bar is open so try to relax and look forward to tomorrow. Be in the reception of B wing at 0800. We will then divide you into groups and outline the training programme.'

'Stand up,' called Buckden immediately after her last utterance. We obeyed, looking at each other as much as the training team as they left the room. I was delighted. I had signed up for this, and I was finally getting started.

Not wanting to come across as antisocial, I made my way to the bar, having seen some of the others from Course Five gravitating towards it. I noticed that a few had gone back to the accommodation block as well. Being one of the last to enter the lounge, the queue for a drink was quite long, so instead of joining it I sat down on my own and sent a message to my girlfriend, letting her know that I had arrived, unpacked and put up her picture. She messaged back quickly and helped me occupy a few minutes until the bar was finally clear of thirsty police recruits. Her own training was going well; she was well into the long process of becoming a doctor. Promising to call her once I had got back to my room, I put my phone back into my pocket and bought myself a drink. A quick look around showed that the smaller groups of the function room had now merged into larger groups at the bar. I chose a group and sat down with them. Although Inspector Preston had kindly suggested we 'relax' and 'look forward to tomorrow', it became clear that not everybody was feeling so calm.

As instructed by the inspector, we were in B wing on time the next morning for a tour of CTC. Our cohort of recruits, garbed in various styles of business wear, were introduced to the administration staff and to those instructors who were not present at yesterday's briefing. With every recruit and instructor in the same room, we were given a simple task.

'Introduce yourselves to us and to each other,' said one of the training team. 'Tell us who you are, where you come from and one interesting fact about yourself.'

It didn't seem unreasonable. These kinds of details would surely become apparent over the next few weeks and months, so an uncomplicated ice-breaking exercise was more than welcome. I heard many interesting facts about previous employment, places visited and unusual pastimes: a hobby locksmith, several special constables and a semi-professional football player; a divorcee looking to start again, and more than a few military veterans who were moving on to their next career. Then came my turn.

'I'm Matt, from Gloucestershire. Used to be in the army. I'm a twin; luckily for him he got all the brains and all the looks so I'm here as a police recruit...'

My attempt at humour was evidently funnier to me than it was to my new colleagues. Either that or my delivery needed more work. With a simple 'thanks for that', the staff moved onto the next introduction while I resolved within myself to make less attempts at humour.

By the end of the introductions I had ascertained that I was the youngest officer on the course by a few years; the next youngest was a twenty-seven-year-old Belfast lad called Ryan. Previously a security officer at Belfast Airport, he had decided to give policing a try.

The exercise served its purpose, and by the time it was finished we had been separated into two courses: 5A and 5B. The students of 5B were destined to deploy to units in Scotland, where law and procedure differed from that in England and Wales, home of the 5A recruits. We were further subdivided into groups that would be assigned to particular units. Sellafield was to receive more than half of the intake, Harwell in Oxfordshire was to take on seven new officers, and two were slated for the East Lothian power station at Torness, while I was destined for Oldbury along with two recruits called John and Mark. The rest were heading to Dounreay at the northern tip of Scotland or to various other power stations in the south of England. Student officers were deployed wherever the constabulary required. Any preferences would be taken into account, but the overriding factor was operational need. I had not indicated any preference so was very lucky to get Oldbury. Not only was it a power station unit, where general policing duties were more common than at the large sites, but also it meant that I would not have to go through the stress of moving across the country like many of the other recruits.

The first two days consisted of 'corporate training'. The content was only loosely related to our future roles and focused more on the UK's nuclear industry as opposed to policing. I found it interesting, but I could understand the frustration of some of the others on the course. Subjects such as the police federation (a type of union for CNC police officers), general health and safety, the effects of radiation, diversity and introductions to the constabulary's structure were covered. I liked it. To my mind, it was important to know the distinction between CNC coppers and regular – Home Office – police officers, which was a product of the respective needs of the nuclear community and the public. I knew enough about the police to realise that responding to crime and terrorism was not just a matter of looking after the cities and towns. Some of the largest crimes in history had involved disruption to infrastructure, not offences against individuals, and these crimes tended to have lasting effects or a global impact. While the damage is impossible to quantify, it certainly reflects the need for multiple law enforcement agencies, including the Ministry of Defence Police, the British Transport Police and various port police services.

Buckden concluded the corporate training by saying, 'After we get you kitted out the police training will begin. It's all been nice and easy so far, but from tomorrow there is a lot to learn and not much time to do it in.'

I looked around and saw that everyone was focussed on the sergeant.

'The CNC has a lot of responsibilities that the Home Office forces either cannot or will not do. Across all our sites there are tens of thousands of workers. Just like anywhere else, they can get into trouble or cause trouble. Thefts, drugs, assaults. Sellafield alone has forty miles of roads. We even have train stations and harbours to cover. You need to know about traffic. You need to know how to support your host forces. Until you know all of that, you cannot go to firearms. So, from tomorrow, prepare to be busy.'

Following these light but revealing first days, we headed to Sellafield, the CNC's principle site and the location of the constabulary equipment stores. Prior to joining we had completed forms with our measurements on them and were now on our way to pick up our uniforms and gear. As the minibuses made their way through the Cumbrian countryside we passed by road signs and place names that were brand new to me, and sounded odd compared to those in Gloucestershire: Beckermet, Bigrigg, Egremont, Seascale. I could imagine that there was probably a correct way of pronouncing these names, but I didn't know it. Coming around a long right-hand curve at a place called Blackbeck, Sellafield came into view for the first time in my life. The site looked huge; it was a town in its own right. No wonder it had always hosted its own police force. The large central tower that used to form part of the generating station dominated the landscape. It wouldn't have looked out of place if the flaming Eye of Sauron from *The Lord of the Rings* sat atop this massive edifice, watching us approach the gates of the domain. As if to add to the atmosphere of our inaugural visit, that morning the bright autumn weather had given way to rain and cloud.

Stepping out of the transport, we made our way to the police stores to collect our uniforms and equipment. I could feel the drops of rain easily penetrating my cheap, machine-washable suit as I waited to sign into the logbook of the equipment store. One of the many site regulations required those who entered building 357 to mark their arrival and departure times. The queue of thirty recruits waiting to complete their entries extended beyond the foyer of the building by some distance. Eventually, though, we were all inside and sitting on school-style PE benches, chatting away and waiting for our turn to collect our kit bags.

I was sitting with John and Mark, my fellow Oldbury assignees, and they quizzed me about Gloucestershire. This was an unknown part of the world as far as they were concerned. A London boy, Mark had never visited many places outside of the city. For my part I had never been to London other than to catch a plane, so we were even on that score. A fellow veteran, Mark had been part of the Household Division and was even based in the city as part of the Blues and Royals; he occasionally went on exercise but always returned to London afterwards. Now in his late thirties, Mark was a short but athletic guy. He looked like a football player who had smoked forty cigarettes a day since he was sixteen, giving his skin a weathered appearance. He had also worked in the prison service but had grown tired of being locked in the same building as the 'lags' under his charge.

'I was thinking about a place not too far from work, but not too close,' he told me. 'Churchdown. You ever heard of it?'

I had heard of it as I had been to school near to the village.

'Yeah, it's close to Cheltenham but it's going to be a long drive to work. You're probably looking at forty minutes or more, depending on the traffic.'

He didn't mind that. I guess the Londoner and the country boy had different ideas of acceptable commuting times.

John, however, wanted to live closer to the site so that he could commute on his bike. His American fiancée was due to join him in the UK once they had tied the knot, and he was leaving the location of the family home up to her. John was thirty-one, and a former paramedic. He had enjoyed his time on the ambulances but decided to become a police officer after spending years working alongside them. Tall and lean, he had dark eyes and a sensible haircut, almost like Ken of Barbie doll fame. But despite his athletic appearance, he hated to run because of shin splints, a condition I had heard of but knew very little about. He was also serious and focussed most of the time, so great to study with but not great for jogging.

'Matthew, Mark, John,' called an instructor from across the wide hall that comprised the CNC stores and firearms training department, 'go into the stores and collect your kit.'

'Luke will be along any second, Staff,' said Mark as we made our way to meet the store clerk and pick up our new equipment.

'Who's Luke?' I asked.

'Matthew, Mark, Luke, and John, right? The Apostles?'

Maybe if there was a recruit called Luke, he would have been destined to be assigned to Oldbury as well.

The equipment store was a two-storey warehouse nestled inside an even bigger warehouse. Larger than a football stadium, B357 was one of many buildings used by the CNC at Sellafield. It also housed the firearms training unit. We were destined to come back here, provided we passed the law and policy aspect of training. There wasn't a single window in the building. It was more like a huge aircraft hangar with a collection of smaller buildings within. When the CNC was a smaller force this had probably been ample infrastructure, but now it seemed overcrowded and congested. There wasn't even a spare inch of floor space as far as I could see. There were offices, breakrooms, classrooms, a shooting range, a PT area and a firearms tactics area as well as the massive equipment store, which was run by the ever-popular Marie. She had been running the stores for several years and had been a part of the UKAEAC. Everyone knew her because she had met everyone from the chief constable down, and everyone liked her because she was good at her job. She handed me a bag with the number 142 stencilled on it under a CNC police emblem. Then she put a printout of the bag's contents on top.

'Check the list against what's in the bag,' she began. 'Check that you have the correct sizes, try it all on 'cos once you've signed for it, then you're wearing it whether it fits or not.'

That sounded good to me. I was about ready to end my involvement in this long and slow day. All of us were carrying bags and boxes containing everything a new police officer could need. The novelty would soon wear off, however, and if everything went well, tomorrow would be the first of many days and many years in uniform.

The following day we were paraded on the gravel track outside of B wing. Buckden and his training team had told us to be on parade that morning, as opposed to coming straight to the classrooms as we had done the previous few days. They wanted to inspect our uniforms, but they had given no instructions on how to get them looking parade ready. The male recruits were wearing their custodian helmets – that emblem of British policing – along with dark-blue NATO-style woollen pullovers and white shirts complete with clip-on tie. Our two female recruits wore their standard-issue bowler hats and had police cravats instead of ties.

The standard of appearance varied between recruits at this early stage. Those with military experience had a reasonable idea of how to wear the brand-new uniforms, which in many cases were still creased from their time in storage and transport back to CTC. They had offered advice to any who sought it. Personally, I found success with a damp pillowcase pressed against the offending creases and ironed. I had been in the laundry room with John the evening before, showing him the difference between good lines and bad lines on a uniform. I had also brought my own boots from home with me, so that I would already have clean and polished footwear. Mark had shined his own boots as well as several other pairs while watching TV all night, which we all agreed was a superb effort.

Whereas most people had sought a little advice in preparing their uniforms, some had not, and they looked like they had slept in them. A guy called Jamie in particular looked like he was wearing a cheap costume for a fancy-dress party. His body shape probably didn't help; even though he was a fit former footballer who had been playing in America, his long legs and short torso did not lend themselves to the standard sizes of clothing. His torso seemed to get wider as it went up, so he looked like a triangle with long legs.

'Good creases, Jamie. Too bad they're all in the wrong fuckin' places,' noted the sergeant.

Another recruit, Robert, was looking distinctly awkward as well, but that was how he looked no matter what he was wearing. On the other hand, Banksy, formerly of the Scots Guards, was the pinnacle of good turnout. His uniform was immaculate and he stood on parade as if he was at Buckingham Palace and not the gravel path outside of the CTC main entrance. When not on parade, Banksy was a cheerful and energetic recruit, destined for Harwell. He seemed to get on with everyone. At first I needed him to repeat what he was saying, as I found it hard to understand his strong Glaswegian accent. The one word I did understand, however, was 'nugget' – his nickname for everyone except the training staff.

Our parade also served the purpose of getting us to look respectable for that afternoon's big event, when we would all be taken into Whitehaven to be sworn in as constables by the local magistrate. This attestation is required for all police officers. The Police Act, which is an Act of Parliament that applies to British police officers, requires all full-time and volunteer police constables to take an oath or declaration. We didn't have to learn this long piece of legislation but were instead handed it on a laminated piece of card.

The magistrate called us forward one at a time while the remainder waited in the gallery (or viewing area) of the court.

'Matthew Okuhara,' she said, sat with a clerk behind her court bench. She wasn't stern or unkind. She obviously knew how important a day it was for some of us. 'When you are ready, Mr Okuhara, please read the oath.'

I looked down to the card that was resting on the prosecutor's desk, where I stood by myself before the magistrate.

'I, Matthew Okuhara of the Civil Nuclear Constabulary, do solemnly and sincerely declare and affirm that I will well and truly serve the queen in the office of constable, with fairness, integrity, diligence and impartiality, upholding fundamental human rights and according equal respect to all people; and that I will, to the best of my power, cause the peace to be kept and preserved and prevent all offences against people and property; and that while I continue to hold the said office I will to the best of my skill and knowledge discharge all the duties thereof faithfully according to law.'

The clerk had been taking notes. The magistrate, who had been smiling throughout all of our attestations, simply said, 'Thank you PC Okuhara.'

She looked at the recruit behind me and called him forward. The process began again, and I took my seat in the gallery.

# 7

We had been in training for a few weeks by now and were getting ready to complete the law and policy aspect of the course. This focussed on general police duties, the kinds of things that officers up and down the UK deal with. Before becoming firearms officers, we had to have a solid understanding of general crime and how to tackle it appropriately. In other words, we had to become beat cops for a few weeks. Even as a specialist police force, we were expected to be capable of assisting our host force counterparts where necessary. Subjects such as public order, assaults and theft were all covered. This aspect of training was easier for some than it was for others, especially those who had previous police experience; several among our cohort had been special constables in the past.[1] Applying classroom knowledge to real-world situations was going to require a lot of practice, so we began to regularly patrol the grounds of Summergrove in hi-vis jackets and police helmets, dealing with incidents prepared by the training team and resolving disputes while pretending to be on a nuclear site.

1 Special constables are unpaid volunteer police officers, with the same authority as full-time professional officers.

The first few weeks moved quickly, partly because we were so busy but also because everything was interesting and new. Not everyone had made it this far, though. Robert had been let go, having got himself arrested after fighting while on a night out back home one weekend. He hadn't seemed violent; in fact he never even came across as *social*, so it was strange that this was his reason for being dismissed. I couldn't remember seeing him in the bar all that often, unless we had organised a revision session, which we occasionally did prior to an exam. Even then his contributions were minimal.

These study sessions were very useful. Studying on your own is one thing, but revising as a group was a different way to review all the new information that we had been absorbing. Each subject seemed to have an expert, or a recruit who grasped the concepts better than the rest of us. The specials (former special constables) had no need to relearn simple legislation such as theft, so would lead the revision sessions on this subject. Other recruits, like Ewan, were excellent at remembering the complex wording of some of the more difficult legislation we were trying to commit to memory. An inspector in the making, for certain.

Along with policing policy and criminal legislation we had to learn about nuclear industry laws and regulations – the kind of information that the territorial police forces[2] are not taught. These various acts and statutory instruments are also not mentioned in any of the handy textbooks that many police officers use as a reference guide. Recruits to the Metropolitan Police or county constabularies have access to tried-and-tested experience and case law. Much of this information is contained in the Blackstone's

2 Territorial police forces make up the vast majority of British police. They are assigned to a geographic area whereas the CNC, BTP and MOD are assigned to a specific industry.

textbook series, which a lot of officers come to know when training or studying for promotion. We did not have that luxury for nuclear laws and the Energy Act. There was only one way to learn about these laws and policies, and that was by coming to CTC.

'Okay then, Banksy,' I said in a study session before the bar was open. 'What is a trans-shipment site?'

Section 56 of the Energy Act defined a lot of the CNC's powers, and this was a tough question on it. Even if you knew what it was, being able to remember how its definition was worded in the legislation was not easy.

'A trans-shipment site,' he began haltingly, 'is a place which a member of the constabulary reasonably believes to be...'

He had to think about this one. Everyone did except for me, as I had it written on a flash card in front of me. Ryan, our Northern Irish security guard, jumped in.

'It's a place where a consignment of nuclear material in transit is trans-shipped or stored; or a place to which a consignment of nuclear material may be brought to be trans-shipped or stored while it is in transit.'

'Superb,' said Banksy as I handed him the flash cards. He would be choosing the next question.

'Right then, you nuggets...' He flicked through some of the cards, trying to find a tricky question, and looked at me. 'How far out to sea does the member of the constabulary have the powers and privileges of a constable?'

'12 miles,' I shot back in a flash.

'You sure?' He was looking at the card and not at me. That made me less sure.

'Yeah. 12 miles.'

'Aye, good guess but it's 12 *nautical* miles.'

He was right. A nautical mile is slightly longer than a statute or 'normal' mile. It was important to get these facts correct, not only as were we in training and likely to be tested on them, but also because the CNC does a lot of work at sea and the law changes depending on the position of the ship in relation to the nearest coastline.

The bar was now open, but we continued our study group for a while longer as we were expecting to have a 'knowledge check' on Friday. The casual name hid the reality of these 'checks', which were in fact exams that we had to pass. Anyone who failed would have to re-sit the test the following Friday. If they failed twice, they would be invited to an interview with Inspector Preston, who would decide whether a third attempt was justified or if alternative action should be taken, such as the individual being back-coursed or removed from training entirely. I had no intention of failing any of these exams.

In addition to these assessments, we also had to pass fitness tests. There were three of these over the first eight weeks of training. Again, unsatisfactory performances would earn an interview with the inspector. We had completed our second fitness test earlier in the week at B357, the only space large enough to accommodate this kind of activity. It took less than an hour for the whole course to be assessed. In the CNC we needed to complete an MSFT (multiple-stage fitness test). This was simply a shuttle run on a 15-metre track, set to a rhythm that gradually increased. We had to reach level 8.9, which was not too ambitious. The angry PTI from the application process had brought us up to level 6.3; an extra couple of minutes running was achievable for all of us. Fortunately he was not running our fitness sessions, which were instead taken by his deputy, a keen fell runner.[3] After our second fitness test we began our PST or self-defence training. Completing this two-day input meant that

3 Fell running is a sport unique to the north of England. Runners race over the many mountains and hills in that part of the UK, known as 'fells'.

we would be able to use physical skills in our patrol scenarios. Up until this point we had only been speaking to 'members of the public' who were in fact instructors or occasionally the admin staff from the CTC offices. Our training was slowly growing in complexity.

The PST instructors were all members of the firearms training department. Sergeant Williamson, who ran the course, was a huge man with a shaved head and muscles all over. Apparently he enjoyed Iron Man challenges, endurance races that involved swimming 2.4 miles, cycling 112 miles and then running a marathon. I was fit and did the occasional half-marathon but was not an Iron Man. As if endurance sports weren't enough, he also owned his own atlas stones, the massive boulders you see on Strongest Man competitions, which he had cut for him by a local quarry. He looked like a henchman from a James Bond movie. Being the size that he was, it occurred to me that he probably didn't need to know any self-defence skills at all, because nobody in their right mind would pick a fight with him. I hoped that he wasn't going to choose me for any demonstrations.

The sarge started with the most basic instruction possible. Holding up a short black rod, he said, 'This is a defensive extendable baton.' He flicked his wrist, instantly extending the weapon. 'I will now take you through the component parts and maintenance issues you might encounter.' I was not sure about what upkeep a metal stick might need. Apparently quite a lot. Here was the end cap, here was the grip. He was pointing to the various parts of the baton. 'This is the shaft, and this is the tip... of the shaft. Don't start playing with it until I tell you to.'

The fitness area that we had been using at B357 had been converted into a dojo for the combat training. John was apparently already an expert martial artist and knew Krav Maga, an Israeli self-defence system. There were other martial artists in the group as well, but I wasn't one of them. We were reminded, though, that in most cases martial arts were sports in which the opponent was following a set of rules. That would not be the case in a physical confrontation while on duty. Of course, fighting was not the only skill taught. In fact, the vast majority of the training was spent going over arrest procedures, physical restraint techniques and how to search people. Over the two days we practised simple methods, and very rarely was it one-on-one training; the odds were always stacked in our favour during these scenarios. By the end of the course we had been issued our handcuffs and extendable batons, now being 'qualified' to carry them. From this point on, they were part of the kit with our notebooks and pens for scenario-based training.

The next day I was working with Rachel, one of the female recruits. We had been told by an instructor while 'on patrol' at the 'Summergrove Nuclear Power Station' to speak to an unknown member of the public who had been hanging around the bike sheds. Rachel was a softly spoken Geordie, blonde and quite petite, but she never seemed to smile. I didn't know all that much about her as she tended to keep herself to herself when not training. She must have been a football fan, though, because she often wore her black-and-white Newcastle United top. Like many of the course members she was destined for Sellafield after graduation. Our scenario was intended to replicate a low-level incident at a large site like Sellafield or Dounreay, where the likelihood of general policing duties was increased by the lack of a host force presence. If support was needed at these sites then it would not arrive quickly, so the CNC had to make up for this shortcoming.

We walked out of B wing and travelled the roughly 20 metres to the bike sheds. Sure enough, we found the man who had been described to us – the instructor had described himself, and subsequently put on an old army jacket to disguise his police uniform.

'Excuse me,' came the soft voice of my companion.

The instructor, who was by now playing 'the bad guy', glared at us and didn't say a word. His mouth tightly shut, he went back to looking at the bikes, lifting them off the ground and spinning their wheels.

'What are you doing?' she continued.

I was two or three steps behind her, having agreed on approach that she would make contact with the subject. Again, no words. Just a glare and back to the bikes.

'Is that your bike?'

I could sense this was going nowhere fast, but didn't interrupt.

'Have you got any ID? Where did you come from? Why are you here?'

The questions kept coming, only to be met with silence. Presumably the training team were looking for something other than a conversation this time. After all, we had just finished the PST course. He took a chain-style bike lock from one of the bicycles and wrapped the end of it around his fist.

'What are you doing?' It was a repeat of her first question, but this time with a note of concern. The lock was hanging down from his clenched fist like a short whip, with the combination lock swinging freely just off the ground. I grabbed my baton and extended it. A simple flick of the wrist is all it took to move the device from the closed to extended position. Rachel did not do the same. She turned and looked at me, confused and a little concerned.

'Put it down and step back,' I said.

The penny had now dropped. We had recently covered the subject of offensive weapons. This included anything that was made, adapted or intended to cause harm. He was an angry man, staring down two probie police officers, and had a chain lock wrapped around his fist. This was definitely an offensive weapon scenario.

Rachel persisted in her questions while I thought about what to do next. I was conscious of the rest of the recruits, some distance behind us, who were watching, close enough to hear but not so close that they interfered with the scenario. Overacting, almost as if in slow motion, the 'intruder' raised his arm as if to swing the chain. I couldn't hit him; not with an actual steel baton. It was only training, after all. However, we had been told that if we felt the need to strike then we should verbally indicate our intention to do so.

'Baton strike!'

That felt awkward to say. A bit like saying 'bang, bang' when training with the army – on those occasions when there were no bullets left to fire, so we were left to pretend.

A scuffle ensued and the reason for the man's tight-lipped demeanour became apparent. Rachel, attempting to restrain the instructor, all of a sudden had a mouthful of chewed-up biscuit spat onto her boots and trouser legs. She shrieked, clearly not expecting it. Nobody had been expecting it. The rest of the course chuckled, partly out of relief that they were not in the firing line but also because it was entertaining to watch recruits struggle through unfamiliar situations. Eventually gaining control, we went through the arrest procedure that we had been taught the day before. I tried to caution[4] our now compliant subject.

'You are under arrest. You do not have to say anything but...'

The words that I used to know so well had been diluted to a memory that I was struggling to grasp. This was certainly going to generate some negative feedback.

---

4 Police officers in England and Wales are required to 'caution' an arrested person either upon making the arrest or as soon as possible thereafter.

'Just shut up or I'll have to write it down.'

That would have to do for now.

The scenario ended. The instructor got to his feet. Sergeant Buckden, who had been making notes throughout, joined him as the rest of the course gathered round to listen to the feedback.

'Right,' he said, smiling broadly. 'How do you think that went?'

*This is not going to be full of praise*, I thought to myself as I dusted off my knees.

# 8

Following a successful eight weeks of general police duties and a well-earned weekend off, the members of Course Five were back at Summergrove preparing for the IFC (initial firearms course). The purpose of the training was to allow the police recruits, upon graduation, to become part of an ARV crew at a nuclear site. With the general policing phase now complete, the firearms phase was looking to develop the recruits to the NPFTC standard (National Police Firearms Training Curriculum), that applied to all armed officers in the UK. Previously the UKAEAC, as a standalone entity, created its own training with a specific focus on the role of protecting nuclear material. The CNC, however, needed to satisfy national guidelines, with the focus of the scenario-based training being relevant to the types of incidents to which we would be deployed. That meant heading back to B357 at Sellafield, where the firearms training facility was located.

Firearms instructors – sometimes known as 'the blue gods', owing to their distinctive firearms coveralls and divine ability to end a recruit's career at the stroke of a pen – had a reputation within the constabulary for being harsh and demanding a lot from the students on the IFC. At the bar, the students of previous courses told us about some of the reasons recruits had been removed from the police entirely. The biggest issue, of course, was an inability to shoot accurately. But there were also incidents of poor safety and even ill-discipline that saw careers over before they had started. I wasn't too worried about the IFC, however, as I had plenty of barrel time in the army. I knew that I would be learning new skills and operating new weapon systems, though.

More concerning to me was the use of force – deadly force – and then being able to justify it. Even if I could justify it to myself and colleagues, being able to articulate my actions in relation to a perceived threat was going to take some work. We knew that this, too, formed a part of the training, and understandably so. The year before, Jean Charles de Menezes, an unarmed passenger on the London Underground, was fatally shot by officers who had mistaken him for a suicide bomber. In 1999, Harry Stanley had been killed by armed police who were responding to an 'Irishman with a gun wrapped in a bag'. In fact, the painter and decorator had been carrying a table leg. I wasn't worried about shooting accurately; I was more concerned about shooting justifiably.

To get onto Sellafield every morning, the twenty-nine of us needed to repeat the process that we had followed when picking up our uniforms and equipment: go to the pass office, pick up the pass, then get on site through a line of traffic that would frequently go all the way back to Whitehaven. This meant a very early start every day if we were to be on time. Winter was fast approaching by now, and we got into the buses while it was still dark outside. There were no windows at B357, and we were expecting to finish in the early evening most days.

We said our goodbyes to daylight for the time being and arrived at the firearms training facility on time. We were immediately divided into groups called syndicates and set about our initial training. As I had expected (following the staff introductions), this involved an overview of the weapons we were going to be handling as well as their component parts and how to clean them.

There were four syndicates, each with two instructors. Additional firearms staff were brought in depending on the activity. My syndicate was made up entirely of officers destined for the power station units: Oldbury, Dungeness, Sizewell and Hartlepool. Our instructors, Chris and Jacko, made it clear that so long as we did 'what we were told, and when we were told', then passing the course would be no problem. They had both been UKAEAC officers for quite some time, with Chris previously having been in the Royal Navy as a helicopter crewman. Although he was very direct and blunt, I liked him. Others were not so keen about his manner. While waiting for the shooting range to become available, he was making idle talk with our syndicate and asked the standard instructor to student question: 'Why did you join the CNC?'

There were various answers. I mentioned that I had always had an interest in the police, and John brought up his paramedic experience and a desire to move his career in a new direction. Adam, a Scottish officer, had three uncles who were in the CNC. Another Scottish officer spoke up with a different answer.

'After this I want to train to be a hostage negotiator.'

Chris shook his head. 'Don't be ridiculous. Don't you know what you're coming into? The only negotiators the CNC have sort out business contracts for uniforms. And sandwiches. And they usually fuck it up.'

The low chuckles after that brief exchange probably killed any ambition for hostage negotiation.

We soon put this talk to the backs of our minds as the range, now cleared of brass cases, was available for our introduction to the Glock 17 pistol.

'As you should know by now,' began Chris, 'this is the Glock 17 pistol. It is called Glock 17, not because it holds seventeen bullets, but because it was the seventeenth patent of that firearms manufacturer.'

We had already had a day of classroom instruction on the pistol, and had been taken through the components, maintenance needs and technical data.

'On this shooting detail you will fire five rounds, under the supervision of an instructor, to a target, 7 metres away. This is for familiarisation purposes only. Afterwards you will strip and clean the weapon before breaking for lunch.'

The range was nothing like I had experienced in the military. They were usually outdoors and nearly always uncomfortable. This was a long, wide annex of B357, with a large soundproof booth at the rear. The consoles inside could control the light levels, activate blue police-style strobes, control a PA system and record a student's activities. There was also an apparatus to rotate the targets that we would be shooting at. 50 metres away, at the other end of the range, were eight targets. Known as Elvis targets, these black-and-white targets featured a distinctly 1970s-looking man holding a pistol, complete with dark glasses, black tie and a quiff of hair. Beyond the target was a thick rubber wall that absorbed the bullet impacts. The smell of cordite and rubber was distinctive and the extractor fans were working non-stop. During a particularly long shooting detail, this would cause the pressure to change inside the building, causing our ears to 'pop', similar to being on an aircraft flying at altitude. Along the concrete floor, lines marked out the distances to the targets. There were also vehicle doors so that the officers of the Road Escort Group (REG) and Tactical Response Group, who regularly deployed on specialist tasks, could practise their shooting skills from their armoured police vehicles. It was all quite impressive, and I felt lucky to be there.

At the 7-metre line, the pistols were laid out with an instructor standing by each one. We made our way down to the guns and went through the loading procedure. These pistols don't have a safety catch, and to begin with this felt a little strange to me. Chris told me to holster, which I did, and he took me through some of his 'top tips' for shooting. I could see that other instructors were doing the same, or at least something similar. Soon enough we were standing in a line, weapons drawn, ear defenders on. We were wearing our 'black kit': black coveralls, body armour and belts with holsters attached.

'In your own time, five rounds to the targets to your front. Carry on.'

Chris had finally delivered the instructions to our syndicate to begin shooting. I took up the slack on the trigger. I squeezed it and eased it back. Bang! Low shot. Around the belt area. At least I hit the target. At 7 metres it was easy to see the 9mm holes that were being produced. Bang. Chest area this time. Good. Bang. Low again. Bang. Low and left. Bang. Good, right in the middle of Elvis. Then there was a pause.

'Has anyone not finished shooting?' Chris's loud voice carried well in this cavernous shooting range. Everybody was done. 'Unload!'

With the weapons safe and holstered, the several instructors advised their students on what they had seen.

'Matt, you're too wound up. You're squeezing the hell out of that gun and shooting low. You can get away with it this close but any further back, then you will miss the target and start shooting lumps out of the range. Tomorrow, we'll work on getting you settled. I thought you were in the army? You should know all this already.'

I was disappointed. It was not a good performance and I could see that the rest of the group had performed better than I had, even those who were picking up a gun for the first time. We had covered the 'principles of accurate shooting' the day before. I had known these anyway, so why wasn't I applying them? He was absolutely right, my grip was too tight and it was causing me to shoot badly. If this didn't get sorted then I would struggle with anything over 10 metres away.

In the student breakroom, we were all discussing the morning's shooting. Fitting twenty-nine recruits into this room was not ideal, especially with the lunches all laid out like a buffet on a table beside the door. Everyone was enjoying the cold finger food, and the shooting. There was very little pressure at the moment, but we knew that we didn't have long to meet the standards required for authorised firearms officers. Wayne, an instructor, had joined us to offer some words of wisdom during the break. He was more than likely there to enjoy some of the food, which he was piling onto a paper plate while explaining that teamwork was essential to pass the tactics phase.

'Remember, there's no "I" in "team",' he concluded as he walked back to his office, full plate in hand, 'but there is in "chicken goujon".' Everyone laughed, but I was thinking about the G36 introduction that would be coming up that afternoon. I did not want to underperform again.

The G36Ks provided for our practice were slightly longer than the G36Cs that we would take on patrol at the power stations, but those at Sellafield, Harwell and Dounreay would be using this K variant of the firearm. Regularly featured in movies because of its angular and futuristic appearance, this semi-automatic gun had been used by the CNC and its predecessor for over ten years. It was lightweight, well balanced and easy to handle – I thought so, at least.

Similar to the morning shoot, we arrived at the range and went through a quick refresher on the gun. The instructors got us to adopt a prone firing position 25 metres from the targets. I had a different instructor with me this time, who was kneeling beside me watching my every move. He wanted to make sure that I was following the correct procedure before I started shooting.

'In your own time, go on.'

We began again. Even with the built-in optics, 25 metres was too far to make out any hits with such a narrow round. I picked a point on Elvis in the centre of his body and slowly fired my five shots. This time the results were much better. The five holes were nicely grouped together and probably would have fit under a 2p coin. I felt better about this. It gave me confidence in the G36 and in myself. I thought that if I could focus more of my efforts on the pistol and not become unduly stressed about the G36, I stood a good chance of meeting the required standards for accuracy. That just left tactics and the Firearms Training Simulator, or FATS.

The FATS was a shooting simulator that could hold two students and two instructors. I had previously experienced something similar, but less complex. The instructors could choose from a range of scenarios and accurately measure the speed and accuracy of our responses. The scenario was projected onto a large screen to simulate being on patrol. Police firearms were wired into the simulator and charged with gas to provide a recoil effect. Everything was controlled from a computer on the instructor's desk. There was also a seating area for visitors or other students to watch the situation and officer in action. It was almost like a live-action video game. After these sessions, the instructors – and often other students – would grill the officer on their actions. 'Why did you do that? What about this? Why didn't you go with another option?' It was a good training tool. The main disadvantage was that the system was American, so understandably all the scenarios were American too. Before long we found ourselves searching US high schools for active shooters and patrolling shopping malls during the Thanksgiving weekend.

The days at firearms training were long and busy. By week three most of us were scoring reasonable results on the range. My old colleague Rachel, however, was not. Her pistol shooting was so unpredictable that even the blue gods had trouble trying to raise her consistency. Another recruit, Alan, was also struggling. Destined for a site in Scotland, he had introduced himself bluntly at the start of the course eight weeks ago, explaining that he was looking to start a second career after the breakdown of his marriage and subsequent divorce. We could all appreciate the sentiments he shared with us. Starting again in a new place, with a new job and new outlook, would seem like a good idea if I was in his position. But I never found out what he did previously in life, and he never said. It seemed that he took this 'new start' very seriously and was keen to improve his lot in life. He never mentioned his personal circumstances again after the initial introductions had been completed.

'Just pretend that the target's your ex-wife, yeah? That'll improve your shooting!' said Mark, jokingly and out of the instructor's earshot. Alan laughed a bit but shook his head, dismissing this informal advice. Of course he wouldn't do such a thing.

His shooting did improve soon afterwards, but Rachel was a different matter. As soon as one problem was diagnosed, then another one would develop: shooting low, shooting high, missing altogether. For now things were looking difficult. Maybe it was her gentle nature, maybe she wasn't into it, or maybe it was her petite frame. It was decided that she might benefit from extra time at the range. If she didn't pass the qualification shoot, however, her police career would be over very soon.

At the opposite end of the B357 shooting range was the skills area. This was where we would practise police firearms tactics such as armed building searches, which was possibly the most important skill for CNC officers at the time. In the event of an armed intrusion into one of the buildings on site, the ARV officers would be required to create an armed cordon and contain the area. The FSU (firearms support unit) would locate the threat and the DET (dynamic entry team) would neutralise it. This is how it would work at large sites with a lot of armed officers available. At a power station, or Harwell, the ARV or TRG officers would be required to conduct the building search and then, if possible, await the arrival of more suitably qualified operators, notably those who live in and around Hereford. However, it was accepted that this might not be possible. For this reason, and because officers were regularly seconded or transferred, all the syndicates regularly took part in scenarios that required us to identify, locate, contain and neutralise threats.

The skills area was the size of a football pitch and made up of temporary structures and walls that could be moved around to create different configurations. It was possible to recreate an office block, a fuel store and a turbine hall among other things. This meant that it was constantly in use, especially by the FSU and DET, who would often practise during their shifts with their own instructors – with us recruits having priority during the day. There was a gantry above the skills area so that instructors could look down into the various rooms for a better view. As with the range, the lights and ambient noise could also be controlled. A turbine hall, for example, is very loud, so when training for an incident that took place there the background track was some suitably loud machinery or, more often than not, loud music.

Forming up as a team of four in the skills area, John, Mark, Knoxy and I prepared to enter a 'workshop' that was supposed to be out of bounds to all staff without relevant access – a highly sensitive area. Someone had seen a worker enter the 'building'. No records on the security system indicated that a member of the site had entered so we had to assume that it was a breach. Knoxy was designated as the team leader. He had previously been a special constable and was also heading to a power station site after graduating. He had two young children back home, and they were beginning to miss their dad. He was missing them too – one of the disadvantages of a long residential course like we were attending. He had tried to instil a little parental discipline over the phone as they had been badly behaved the night before. He was heading home at the weekend and looking forward to seeing them.

'One, ready,' John began.

'Three ready, four ready,' we reported to Knoxy, who was number two, and began our entry.

We knew that the building would only have one subject inside. We did not know, however, who it was, what their intentions were or whether they were armed. We also knew that we could make an arrest when we located them as they were trespassing on a protected site – a serious offence under the Organised Crime and Police Act, which had come into force when the CNC was created.

John was at the front, and we followed the left wall inside the building until we came to a closed door. We were now well practised on room entries and barely needed a word to get set. While Mark and John provided cover that looked up and down the corridor, Knoxy and I made our way in. We surveyed the room, G36s at the ready. Nothing. A workshop office, complete with furniture and spent blank bullet cases on the floor from the previous group's scenario.

'Room clear!' I shouted, and we stacked up again to move further into the building.

It was not unknown for the staff to have us clear fifteen or twenty rooms before letting us know it was a false alarm. We all hoped it wasn't one of those scenarios, although they were good for repeatedly training room-entry skills in quick succession. When you are on the door and about to go in, even in training, you need to focus and stick to well-rehearsed techniques. The additional stress of being under assessment added to the strange mix of excitement that most of us felt even after 100 or more room entries. We carried on, room after room, occasionally changing formation to allow each recruit to experience different positions in the team.

We came to a wide set of industrial double doors. No problem: we knew what to do and got into formation. Knoxy and I went in. 'Armed police!' We called in unison as we saw a man in an olive-green military jacket and BNFL hard hat. He was holding a sledgehammer in one hand and a red pass in the other. That was a big hint. A red security pass is only for visitors, and they need to be escorted at all times. We had our man. But he also had a sledgehammer for some reason.

'Armed police,' Knoxy declared again. 'Put the hammer down. Put the pass down. Do it slowly.'

The intruder dropped the pass as I pushed further to the left of the room, trying to get a better look at him and the area immediately around him.

'Put the hammer DOWN!' Knoxy repeated. No response.

The man looked at me and started talking, rapidly and without pausing for breath. 'Don't you point those fucking guns at me,' he shouted. 'Who the fuck do you think you are? What you gonna do, shoot me?! Come on then! Do it! Shoot me!' He was still shaking his sledgehammer but hadn't raised it.

There were a few metres between us and the man. Maybe we could talk him down.

More noise from him. 'Where's that fucker? I'm gonna fuckin' kill him!'

He obviously had a story to tell, but we weren't prepared to listen until the hammer was on the floor, or at least until he was less agitated.

'Stop being naughty!' Knoxy ordered him.

A pause. Then laughter. Simon, the instructor who was playing the role of the intruder, lifted his mask. 'Stop what?!' he asked, grinning widely.

Chris, who had been following us through the building, was similarly amused. 'Is that something we taught you?!'

The exercise was over, and we had done a good job up until then. As we made our way out of the skills area and back to the locker room, Knoxy explained.

'Well, it's just I was trying to get my kids to behave last night and my Mrs called me at lunch. Anyway it worked, right? He stopped being naughty.'

It was impossible to argue with that.

As the scenarios became more complex, we began dealing with multiple subjects and firefights that simulated the terrorist threat to the nuclear industry. Despite the variations in the scenarios, the criteria for deploying armed officers always remained the same, as did the objectives of the deployment. One afternoon, after a particularly difficult morning of room searches and containment exercises, we were called into the gym where we had practised our PST skills a couple of months earlier. Set out on a long table were several deactivated weapons and former UKAEAC firearms.

'We're going to do some weapon familiarisation,' announced Jacko, Chris's number two. He was a powerful-looking guy whose hands seemed too big to hold anything smaller than a shotgun. He was a keen biker and we often saw him walking into B357 wearing his leathers and boots. He spoke with a broad Cumbrian accent but originally came from Malta, just before it gained its independence from the UK. He often took on an acting sergeant role but was reluctant to transfer out of the firearms training unit and end up back on shift as one of the section skippers, despite being qualified to do so.

'It might fall to you guys to make safe any firearms that you may recover post incident, or upon request from another constabulary, so what we're going to do is, go through this lot and look at the similarities,' he explained. 'Then we will look at some particular issues that some of these guns have.'

Reaching for a Heckler & Koch G3 rifle, he began his presentation. 'These used to be carried by the Marine Escort Group, and we still have dozens of them in the armoury…'

I could see that this was going to be a long presentation, and settled into the bench as best I could, waiting for the opportunity to get a closer look at some of the guns I had not seen before.

All too soon we arrived at test week, the last week of training for most of the recruits on our course. This final few days of firearms training was going to be nothing but assessments. If we weren't being scrutinised in the skills area or on the range, we were going to be waiting. There was a lot to do. We had to prove our competence on the Glock 17 and G36. We had to demonstrate that we could carry out an armed response in various situations, and we had to justify the use of deadly force in the FATS simulator. There would be four days of tests before a return to the CTC for the verdict. We did not know the order in which we would take these tests.

I was hoping to get the shooting out of the way first; I had become consistent with my accuracy but still knew that the biggest cause of failure at this stage was poor shooting. I was in luck, for my syndicate was going to start off on the range. We were allowed one practice run of the test before going through with the formal assessment. This time we had new and unfamiliar instructors. I guessed they were from the DET or FSU. We were lined up for pistol practice and heard the familiar words of command. No problems so far. After clearing the spent bullet cases from the range we recapped on the G36. Finally, we went on to the actual test. One thing was going to be slightly different this time, however. Instead of inspecting the targets ourselves, after each shooting detail an instructor would go forward and record the scores. A previous course had seen officers not patching the targets out properly on a test – trying to inflate their final scores or secure a pass on an element in which they lacked confidence.

We started with a 7-metre shoot. Then we went back to 10 metres, 15 metres and finally 20 metres. This was apparently the detail that caused the most problems for recruits. At 20 metres, the officer had to fire a single shot at the target over several exposures. That meant the target would repeatedly appear, hold for a few seconds and then disappear. To pass, the officer had to land at least half of their shots on target. If they did not, regardless of their performance on the rest of the test, they would fail. They would be able to retake the shoot at some point in the week, after they had received some coaching or additional instruction. A second failure would mean removal from the course.

During the last shoot, Chris, who was working the target console, came out of the booth and tapped me on the shoulder. 'You're dipping your head, Matt. You're going to mess up your sight picture.' He patted my neck, indicating that I had been bobbing my head around instead of holding a good shooting position. I hadn't even realised; I was just thinking about the upcoming 20-metre serial. He was back in the booth before the instructors had finished counting the scores. Coaching wasn't allowed during a test. No doubt if questioned he would say he was addressing a safety issue such as ear defenders not being worn properly, but I appreciated him taking just a few seconds to tell me.

The G36 shoot also started close to the targets and worked back to the end of the range. The initial shots were fired from 10 metres away, and we progressed towards the 50-metre point and erected barricades along the way. This assessment was much easier in my mind, and I was hoping that all my shots would land on target. The results would be announced once everyone on the IFC had been through the assessment.

We cleaned up the bullet cases on the range and made our way to the gun-cleaning room.

'How do you think you did?' I asked, looking at Mark but directing my question to anyone that wanted to answer.

John was typically cool about it. 'Probably did okay,' he said.

'Yeah,' Mark added. 'On those close shoots you can see the holes on the target but further back, I dunno. Probably it's okay.'

I felt the same. I felt comfortable with the shooting, as if nothing had gone wrong. Ross had unholstered his Glock and began to clean it.

'You're a bit quiet, Ross,' added Mark.

We found out why later that day. You can often tell when you aren't shooting accurately. Ross had failed on the 20-metre test, along with Rachel and Banksy. They were going to have some extra instruction that afternoon, and would be shooting first thing the following morning. If they failed again, they would not be taking part in any of the other assessments. We shared their anxiety – it could have been any of us.

The next days were spent in the skills area. Starting with a 'single-officer emergency entry', there was going to be a lot of waiting around as each member of the syndicate would need at least fifteen or twenty minutes to go through the assessment. When my time came, Sgt Williamson, who had taken us through PST several weeks before, was going to follow my progress through the corridors and rooms. It was a 'textbook' CNC scenario in which, as the only officer available, I had to respond to sounds of gunfire reported from an office complex. All the staff were accounted for, and a disgruntled former employee was inside, looking for the manager who had sacked him.

'Any questions?' he asked.

I loaded and made ready. There was a reassuring click as the working parts took the top round from the magazine and held it inside the firing mechanism.

'Armed police!' I called from outside the offices. 'Anyone in the building, make yourselves known!'

At that time, the advice was to give the subject an opportunity to engage in some dialogue where possible – and not to treat shoot-to-kill as the only option. I heard nothing and proceeded inside. Room after room with the sarge and his clipboard following me. I was starting to get warm. Surely I would make contact soon; I felt I had been in there for ages. I wondered if I had been going in a circle by accident. Entering a locker room, I saw the suspect holding a shotgun in his right hand.

'Armed police, put the gun down!'

The gun didn't go down. It started coming up. Bang. I fired. The blank ammunition was only slightly quieter than the real thing. He played dead or seriously injured.

Identified, located, contained and neutralised, I thought to myself, still covering the body with my G36. But there was still more to do. I went into the arrest procedure. Even with the suspect in this state, we were expected to make a lawful arrest before rendering first aid. I was prompted to move on to the next room, and the next and the next.

'End-ex,' Williamson finally said. 'Good, good. No problems.' That was encouraging to hear.

I took off my eye protection and ear defenders and made my way out of the training area. I noticed I was sweating quite a lot, and the cold of the massive building was starting to make me shiver. *Nothing a cup of tea won't solve. I've earnt a brew*, I concluded. The instructors stayed behind to talk and rearrange the building for the next assessment while I joined the others in the breakroom. They had already been through, and we talked about our different experiences in the skills area. The officers waiting to be tested were in a different room.

With so many assessments and tests, unless there was a failure, feedback wasn't due until the last day of the course. But everyone seemed to be in good spirits. Banksy and Ross had rejoined the course as well, having passed their reshoots. They would soon be going through the room entries and rejoining us for two-officer and four-officer building searches. Rachel, however, needed further development and was due to meet with the CFI, or chief firearms instructor, to discuss her options as an AFO in training.

In comparison to the emergency entry assessment, the two-officer and four-officer scenarios were less stressful – for me at least. With all the practice we had done, they were almost enjoyable. While waiting for the two-officer building searches we also undertook our FATS tests in pairs, one officer armed with a Glock 17 and the other with a G36. It seemed that every scenario was the same for every pair of recruits: an 'active shooter' at a shopping centre. I could see the names of unfamiliar shops in the background. JCPenney and Pier 1 Imports were not brands that I knew. Unlike the building searches, though, the FATS did not require us to find the incident. In less than a minute I, along with John, engaged a well-armed maniac somewhere in middle America.

The instructor paused the simulator as soon as John had fired. He had the G36, whereas I had the pistol. We knew what was coming. The simulation rewound for five or six seconds. A green mark danced around the screen, indicating John's aim while he acquired his target, as well as a blue dot showing my aim. The simulator was playing in slow motion and the green mark turned red to indicate the position of John's gun when he had fired. Dead centre. Good shot.

'Why didn't you fire?' This question was directed at me.

Aside from justifying actions, we were also expected to justify a lack of action, if that was thought to be relevant. In this case, for me, it was simple.

'I didn't have a good sight picture and didn't want a stray round going into any of those shops,' I said.

The 'backdrop' is important when shooting. When trying to pick up a good sight picture, you also need to be aware of what is beyond the target. What if you miss? What if the shot passes through the target? It could hit an innocent bystander, damage property or, in the case of a nuclear power station, cause a catastrophic release of chemicals and gases. My reason for not acting was taken on board and noted down. John's reasons for engaging were also written down. More questions. More notes. The scenario was less than two minutes long, but the debriefing took a lot longer. The whole time we stood with our guns still in hand, as if we had just been in action and were still at the scene.

The FATS screen still displayed the portly shooter, with a red mark in the centre of his chest.

'Did you need to shoot? Who was he threatening? Did you feel you were in danger? Did you feel others were in danger?' It was a robust line of questioning. We gave an account of our actions for what seemed like an age. Eventually, we were permitted to leave the simulation.

'ARMED POLICE! ANYONE IN THE BUILDING MAKE YOURSELVES KNOWN!'

That was the loudest I had ever heard Knoxy say anything over the past months. He had been designated as the team leader for our four-officer scenario. I was at the front and felt him tap my shoulder, indicating that I should move forward to breach the threshold from outside B357 into the skills area. It was cold outside but I felt warm – a mix of anxiety and physical exertion from the day's events. Entering the mock-up of a turbine hall, it was no warmer. I knew that the team would be following closely behind and had my G36 raised, covering the area directly in front of me. I was aware that this scenario would likely be dynamic to say the least. We had been told of a pair of intruders on the site who were armed and intent on damaging the generating gear. Possibly there was an insider as well. The firearms training team, who were acting as the armed intruders, had even come up with a name for their terrorist group, and some stated aims.

With no time to await assistance from other units or backup from the military, it was decided that the available CNC resources would deploy in order to locate the threat and take action. With plenty of possible areas for cover, the test area was the most complex I had ever experienced. I supposed that a real turbine hall would be the same – although at this point I was yet to see one in real life. Inside the skills area was a structure the size of a two-storey house. It had no doors or windows. Running in and out of the structure were pipes big enough for a man to crawl into, and insulated cables as thick as oil drums.

Immediately after we had started to enter the area, a camouflaged figure in dirty blue jeans broke cover and ran into a series of offices attached to the hall. He was up and gone in less than a couple of seconds. We could hear him shouting something but over the simulated noise of the turbines it was impossible to understand him. Was he in there with someone, or was the second person still hidden among the machinery and equipment?

Securing our immediate area, John and I took up covering arcs to allow Knoxy to come up with a plan. We were lucky he had been designated as the team leader, as he seemed to naturally fit the role. He decided to search the turbine hall first.

'Those offices are a dead end,' he reasoned. 'If they want to come out, they have to come through us.' It was sound reasoning.

Leaving John and I to cover the doorway, he issued another challenge – this time in the direction of the offices. No reply. Taking Mark with him, he performed a two-person search of the turbine hall. Searching around and over the machinery, they efficiently covered the entire area. No result. That meant we would need to form up and move into the offices.

I was number one again. I liked being number one. I felt as though it suited me, and I had performed a similar role during FIBUA operations in Basra, Iraq. The offices were the same ones that we had used for our emergency entry searches in earlier tests. But we knew that the layout could easily be changed so that we would never be completely familiar with the layout of the area. Rounding an L-shaped corner, we found ourselves in a long corridor, like what might be seen in a hotel. This was going to be hard work. Clearing each room and resetting to move towards the threat over and over again was going to be physically demanding.

I doubted that the people we were looking for were in the first room. But I was also prepared to assume that there were people in every room. Site workers? Hostages? The insider? We knew that the area had been evacuated but that not everyone was yet accounted for.

Making our way down the corridor after clearing four offices, we prepared to start over. Before we had the chance, a door at the very end of the corridor swung open. Two figures stood in the doorway. Clad in camouflage jackets and jeans, they were also wearing balaclavas and toting AK47s. They looked similar to the stereotypical IRA gunman from the Troubles. One of them began to crouch while the other, directly behind him, started to raise his weapon into his shoulder. I didn't have time to issue a challenge. Bang! I fired. From behind me, shots at the same time. The men fell. I kept them covered with my weapon. We could not discount the possibility that there were people still in the building, hostile or otherwise. We also needed to provide a duty of care to the people we had just shot. As this was a policing operation, we needed to secure the suspects, preferably alive. A threat had been neutralised, but we were not out of the woods yet.

'END-EX,' came the call from the gantry above. Looking down on us was the chief firearms instructor, and several blue gods that I had seen but could not put a name to. They must have been watching us the whole time and running the assessment. Getting up and removing their balaclavas, Chris and Jacko made their way out of the area. There was no hint as to whether we had done a good job or a bad job. We would soon be debriefed, and only then would we find out.

The bar was busy that night. Josh, the barman, had anticipated this end-of-training event and brought his younger brother in to help ease congestion at the bar. He had seen more than his share of CNC recruits come through. He probably knew as much about the law as we did, having been able to pick up on the content of every revision group for the past few years. But he was happier pulling pints than pulling night shifts. Everyone had come for a drink, and morale was high. Everyone felt confident. Other than Rachel, nobody had been spoken to 'in private' and nobody had to do any further retests. Rachel seemed pleased with the outcome of her meeting; she was to attend the next firearms course and start her AFO training again, ironing out any shortcomings on the range. I think that was the only time I saw her smile, and the only time I saw her in the Moon Bar.

The atmosphere was jovial. The Sellafield officers were in a large group and having a great time. The Scottish officers were in a smaller group, and the Oldbury crew were in a trio. We each had a cold beer and were sat near the TV, which was playing Tom Cruise's blockbuster *The Last Samurai*.

'Well done boys,' said Mark, raising his glass.

'Yeah. Nice one. Cheers,' John replied.

We hadn't had all our feedback yet, but the feeling was that we had made it. After fourteen weeks of training, we would finally get our end-of-course assessment results the following day. Graduation was on Sunday, and most officers would then get four days off before reporting for duty at their units. Those going to power stations were not done yet, as they had a week-long familiarisation course at Hartlepool nuclear power station. But this wasn't on our minds at all. Tomorrow, we could get our results, enjoy an early finish and head home for the weekend. We had earned a good rest.

# 9

I had been at Oldbury for around six months and things had already started to change quite a lot. For starters, we finally had a proper police station. The disused visitor centre had been converted to a decent-sized base of operations for our growing contingent of officers. It was just outside the power station boundary, making it easier to conduct patrols and monitor the area. That wasn't the only bonus as we now had a kitchen, gym and showers. We found ourselves with a new boss as well. Aled, the station sergeant, had been replaced by an inspector who had transferred in from Gloucestershire Constabulary.

The 'governor',[1] whom we called Mo, was an odd guy. Tall and thin and pale, he often brought up inappropriate topics such as his daughters featuring in 'gentleman's' magazines as they tried to start modelling careers, or how he was totally anti-religion and how anyone who didn't agree with him was deluded. But having said all that, he was a good administrator and experienced in

1 The UKAEAC and CNC inherited the term 'governor' to refer to police inspectors and chief inspectors from the many London-based police officers that transferred into the UKAEA and CNC.

running multiple shifts. With seventeen officers now at the station and more on the way, there was a lot of administration to be seen to and he had plenty of experience of that. Along with Mo, we had received a batch of other transferee officers to bring us up to four officers on each team, as well as an OPU deputy commander, a sergeant who had transferred in from the Metropolitan Police Service. All in all, the look and feel of our policing team was completely different from a few months ago. We all thought it was a shame that Aled was back on shift and not running the show, but B shift were happy to have him as their section sergeant.

I was still on C shift. JT had been poached by the firearms department to begin his training as a blue god, and Honking John had left to join the Met – which had been his long-term goal. In their place another transferee sergeant, Mack, had joined us from Gwent Police. That meant our team of four officers was now Mack, Dave, Cal and myself. It was a good crew. However, despite the influx of new officers, I was still the junior copper. In terms of service I was junior to everyone else, and in terms of age I was still the youngest.

Dave and I were on a patrol on a Sunday morning. I was due to return to CTC for a stage five review – essentially a twelve-month progress check.

'Easy week,' Dave assured me. 'Nothing to it. Basically five nights out in a row with some scenario-based training in between.' That sounded promising. 'You've already covered everything. It's a long drive and a boring week, but you have to do it. And besides, now Gav isn't acting sergeant on B, he wants the overtime from covering your shifts.'

He chuckled. Gav loved his overtime. He was the first person anyone went to if they had plans and needed someone to cover their shift for them. He was constantly checking in with the deputy boss, Beegee, to see if he could get an extra shift. We had no idea whether he was saving up for something or just loved the job so much that he wanted to be on duty every day. I had worked with Gav a few times when Cal and Dave had been away on courses. I had found out that he was a big football fan, and loved to watch Bristol City. He'd even taken me to a game one evening. I'd never been to a 'proper' match before and usually went to see my stepbrother play in the semi-professional leagues instead. He also told me about his job in the RAF, guarding nuclear material at MOD sites, and his life before the CNC. He was basically doing the same job now, with a different badge on his hat.

Our patrol had just exited Oldbury village and was going up Oldbury Hill, which overlooked the whole Severn Valley. The plan was to drive up to Thornbury and monitor the main approaches to the power station while Mack and his crewmate waited on site at the other end. Dave was driving and took us over the crest of the hill and St Arilda's church. This place of worship had sat on the hill for hundreds of years and it was easy to see why. Even on a rainy day, the views were magical. The parish was named after a saint from Oldbury who had been killed in the sixth century. Her death had caused the water in a nearby spring to run red. Mo probably would have suggested it was a fungal problem as opposed to a miracle.

'You recognise that car?' asked Dave, reducing us to a crawl as we started to go back down the hill. On the left side of the ARV, we could see a path that went back down the opposite side of Oldbury Hill. A red Volvo was up to its number plate in mud, having clearly got stuck trying to drive up the track.

'Nah, not familiar.'

The only nearby houses were large stately homes and farms. Neither of us could recall anyone driving a red Volvo. He stopped the car to let Mack know he was going to have a quick look.

'Oscar Yankee One from Oscar Yankee Two,' he began. 'Yeah, we've got what appears to be an abandoned vehicle on Oldbury Hill. Doesn't seem local. We're going to check it out.' It wasn't entirely necessary to report in to the other team, but Mack liked to keep a firm eye on his shift and was stricter than the other sergeants when it came to procedure. Despite only being in the CNC for a few months, he came across like he was from the UKAEA old guard.

'All received,' came the reply. 'We'll monitor your channels.'

Dave picked up the vehicle radio handset to speak to Avon and Somerset. He wanted to run a PNC[2] check to get some details about the car.

'Matt, open a channel with Harwell, let them know where we are and what we're doing.' I had already sent them a message and was waiting for a reply.

'Yeah, let's go and have a look,' I replied. We opened the ARV doors and almost immediately the passenger doors of the Volvo opened as well. Two people emerged and ran back down the hill while another two burst out and ran to the right. There was no way we could catch the pair that were running and sliding down the hill, but the other pair had run into a field, and we knew this field had only one way in and one way out.

'Matt, bring the motor up to the gate,' Dave said as he threw me the keys to the ARV. I caught them and started the car. As I put it in gear I could hear he was talking on the radio to the host force. Mack would be listening in as well, so I didn't bother getting in touch with him. The ARV radio was tuned into the Avon and Somerset channel and I could hear Dave speaking to Uniform X-ray, the callsign for the local control room. The PNC came back as a stolen vehicle. The previous evening it had been stolen from a Bristol University medical student from Vietnam who was staying in nearby Clifton. The registered name and address gave us good

2 PNC, the Police National Computer, holds vehicle, criminal and property records as well as other relevant pieces of information.

information to begin an investigation that we could pass onto our host force colleagues: if the 'owner' of the vehicle was not called and could not pronounce the name of the owner (which was not easy to spell or say), then we would be on to a good start.

I parked our vehicle across the gate to the field. Recalling the subjects that I had covered in training, I thought about trespass and TWOC (taking without consent, usually pronounced 'twock'). I couldn't remember the entire legislation as we had only briefly looked at it, but I remembered it was a little different to theft. To prove the crime of theft, you needed to demonstrate that the accused 'dishonestly appropriated' property with the 'intention of permanently depriving' the other party of it. TWOC did not require the 'permanent' part of that definition but it also only applied to vehicles. We also knew the car was reported as stolen, so any 'dishonest' element of the offence was almost certain. But we also needed to hand this over to the host force as soon as possible. Even though we were police officers and responding to crime in the community, we had to return to our core task as soon as we could. The CNC duty inspector would be expecting an update as soon as we could get one to them.

I heard Dave updating Uniform X-ray while I reported to our own people at Harwell. He was standing at the gate to the field and pointed at the sky. 'Nine Nine was relocating and heard us whilst on his way to Cheltenham,' he said while trying not to grin, and returning his gaze to the two figures now squatting under a substantial hedgerow at the far end of the field. Nine Nine was the helicopter shared by Avon and Somerset and Gloucestershire Police. By sheer coincidence, it was in the area and was listening to the Uniform X-ray frequency while Dave reported to the host force.

'Two armed officers, and a helicopter,' I thought to myself. That's a lot of resources for a countryside TWOC on a Sunday morning.

As Civil Nuclear Constabulary officers, we did not get nearly as many general policing tasks as the other officers in the UK, but our jobs very different in that regard. That is not to say that the CNC never takes part in general policing duties. Most officers deployed to the power station sites will find themselves assisting their host force counterparts at some point. Often this is because policing resources in the areas around the power stations are scarce, but also, by virtue of being on patrol, the ARVs of the CNC regularly come into contact with the public and incidents that cannot be ignored.

'Oscar Yankee Two from Uniform X-ray Nine Nine.' The eye in the sky was summoning us. 'Two possibles moving between your location and Moor Farm.' Moor Farm was the largest farm in the area and a prominent feature on the local maps so we knew exactly where he meant. However, there was no way we could get to them quickly from where we were, as the red Volvo was blocking the track and preventing us from going down the hill to look for them. Besides, if we went looking for those two 'possibles', then what about the two 'definites' we had in the field?

Around the size of half a football pitch, the field was one of a number in a smallholding owned by a family that lived near London. They usually only visited during holidays and had a groundskeeper visit regularly to keep their estate tidy. Surrounded by thick hedgerows and one wide, rusty gate, the field was used for their family's horses and trailers, which often accompanied them. It was a matter of either waiting for the pair to come out or going in and bringing them out. So that we could return to our core task as soon as possible, we opted for the latter. It wouldn't be long before the host force arrived on scene, and it would probably appear less than professional to be providing armed protection to the gate of a field where two kids were hiding. Besides, pressure was coming from our bosses to return to the business of nuclear security as soon as we could; they had to report in detail to the regulators about CNC resources being used for general policing duties.

Making our way across the paddock, a couple of teenage girls emerged from the far end before we had even got halfway. Wide-eyed and pale, their liberally applied makeup failed to conceal their emotions or frustrations. Maybe it was their first run-in with the law, or maybe they were annoyed at the prospect of being taken to Bristol nick courtesy of Avon and Somerset Police. One tripped and Dave helped her out of the muddy ground and brought her back towards the police car at the entrance to the field while I guided the other. They definitely weren't dressed for the countryside; low-rise jeans and wedge heels are certainly not ideal attire for soft agricultural ground on a wet weekend morning. Maybe during a summer fête lower down the hill it would have been okay, but today was a wet, sodden morning and their shoes sank into the soft earth with every step. The likelihood of them running anywhere seemed very low. They had nowhere to run to anyway; there was only one road.

We arrived on the track and Dave cautioned them. 'You do not have to say anything but it may harm your defence if you do not mention when questioned on something you may later rely on in court. Anything you do say may be given in evidence.'

That started a dialogue. 'Where the fuck are they those fucking wankers?!' Clearly upset about being abandoned by their boyfriends, these two were more than a little irate and not interested in talking to either me or Dave. They also seemed keen to apportion the blame to the two runners. They were talking to each other and talking over each other to such an extent that we had to separate them. It was impossible to get a coherent account of anything. I went with my detainee to the rear of the ARV, and Dave remained near the gate. We could write their accounts up later, once back at the station. I noticed that my charge looked too young to be out all night. But then again, I looked too young to be a police officer. I repeated the caution that she had heard only a few moments ago before she gave me her details and those of her accomplice. Surprisingly, she was very cooperative. In her defence, she claimed she did not know that the car was stolen and said the driver was a friend of her boyfriend. I tried to piece together a story that I could pass on to Avon and Somerset. It would be up to them how to proceed, and they would not appreciate it if we just said, 'Arrested these two, see you later.' They would also need us both to supply them with statements so that, if the case moved forward, they would have evidence they could use to secure a good result.

Nine Nine, who was only repositioning when he heard us, was carrying minimal fuel and had long since flown to Staverton Airport in Cheltenham. Uniform X-ray had sent transport for our two teenage detainees, and Mack was being updated on the situation. He wanted us back on site ASAP so that he and his crewmate could begin a search of the area during their scheduled external patrol. There was no reason that their patrol could not work in a little community policing while providing security to the power station. As a transferee, this was one of the reasons Mack and the other Home Office police were brought in; their county policing experience was something that the CNC lacked. Also, the ARV that Mack was using had a prisoner cage and was much more suitable for general policing work than our vehicle. It made sense to head back and return to the core tasks we had been set for the shift.

Dave looked at me a little disappointedly. 'You know, Mack will handle it from here… and we'll be back on site, drinking tea and writing up our statements.'

I understood why he was frustrated. He enjoyed doing general policing and had spoken about transferring to a different force on more than one occasion. However, he was reluctant to move from his home and family for the sake of his own goals. A transfer would normally require undertaking all of basic training again, as

CNC officers do not cover it thoroughly enough compared to our counterparts, instead focusing on industry-specific laws and firearms training. The thought of going through training once more and starting his career all over again put him off the idea for now.

'I don't think it will be too hard to find a couple of chavs wandering around here though,' he added. 'There isn't even a bus to take them home, so unless they feel like swimming across the Severn, then it's not going to be a difficult search. Nah, probably better to head back and do an internal patrol, statements and lunch.'

The transport for our detainees arrived. The Avon and Somerset van was crewed by two local officers from Thornbury, one of whom was a woman. This was ideal, as the two girls needed to be searched properly before being taken into custody. We had performed a basic search after their arrest but knew from experience that they would be searched again, and more thoroughly.

'You not cuffing them?' said the male officer as he drew his own set of handcuffs from his belt.

'Nope,' said Dave. 'It didn't seem necessary and besides, we never get our cuffs back once you've taken them away.' Dave was referring to an arrest he had made with Mark and Gav the month before, just outside of the power station. During that incident, three young men were arrested for possession of drugs. Stupidly, or unluckily, this hapless trio had attempted to divide their stash of suspicious white powder and cash away from prying eyes in town, at a road near the main approach road to the power station. Even more unlucky for them was that they chose to go through with this transaction during our shift changeover, meaning eight officers were available to investigate their unexpected arrival and they were all within walking distance of their beaten-up old Fiat. Even the guard force got a look in, sending out their truck to shut the road exiting the site and recording the whole incident using the many cameras around the site perimeter. It seemed that Dave, more than the other officers at Oldbury, was somehow able to attract policing tasks to the area.

Arriving back at the station, it was easy enough to put together our statements and send them over to the host force. We also needed to put together a GPR (general police report), to send to our own hierarchy that covered the basic details of our incident. Similar to a police statement, the GPR eventually found its way to the divisional commander's office; he would submit regular updates to the governmental department that oversees the CNC on police activities within the entire area. All the details had been recorded in our pocket notebooks, and besides, the incident was not overly complex, meaning that after one cup of tea the paperwork was looking almost complete.

With that administration out of the way, we now had time to conduct an internal patrol and check in with the security team at the reactors while Mack and his crewmate were outside the wire. As it was an immensely complex and secure building, a regular police presence was not always necessary; the physical security measures alone would prevent almost any incursion. However, as any diligent beat cop would know, it was a good idea to check in with the residents of that unique location from time to time.

*Right:* 1. NBC training with the army. At this point I had never heard of the CNC or UKAEA and was considering a career in Gloucestershire Constabulary after gaining some life experience. (Author's collection)

*Below right:* 2. Before a patrol in Iraq. My role in the conflict enhanced my interest in policing and security. (Author's collection)

*Below:* 3. Sellafield as seen from the air. With over 40 miles of roads, tens of thousands of workers and countless buildings, Sellafield is the largest nuclear facility in Europe. It also hosts the largest CNC policing unit. (Courtesy of Sellafield Ltd)

*Above:* 4. Oldbury-on-Severn Power Station. Completed in 1967, officers were not deployed to this location until 2005, following the creation of the CNC. (Courtesy of Brian Clifford)

*Above left:* 5. On patrol at Oscar Yankee. Nuclear power station armed response teams, known as 'support units' regularly patrol both inside and outside of the facility. (Author's collection)

*Below left:* 6. On patrol. A final shot before heading to Escort Operations. (Author's collection)

7. Loading ammunition and supplies onto one of the international escort vessels. (Author's collection)

8. Training. The CNC trains to the same standard and uses many of the same techniques employed by the officers that make up the UK police service. (Crown Copyright/ National Archives)

9. M134 minigun – used for defence aboard the escort vessels. (Crown Copyright/National Archives)

*Above:* 10. General-purpose machine gun (GMPG), mounted for defence. (Crown Copyright/National Archives)

*Left:* 11. The *Pacific Heron* and *Pacific Egret*. Both of these vessels have been used to move MOX fuel around the world. (Courtesy of PNTL Ltd)

*Below left:* 12. Loading the fuel. Using specially constructed containers, the loading process can take several hours to complete. (Courtesy of PNTL Ltd)

13. Counter-terrorism exercises. Officers regularly train with other security agencies and militaries. (Crown Copyright)

14. Road escort officers conducting a briefing in between training exercises. (Author's collection)

*Above:* 15. A C130 Hercules at Wick Airport, similar to the type used to carry out a short-notice cargo move to the USA. (Courtesy of Highland and Island Airports Ltd)

*Left:* 16. Firefighter training as part of the escort group pre-embarkation package. (Author's collection)

17. The *Atlantic Osprey* arriving in German territorial waters, escorted by the *Wasserschutzpolizei.* (Courtesy of Fred Dott/Greenpeace Media)

18. The *Atlantic Osprey* coming into Nordenham, with German escort and protesters surrounding the vessel. (Courtesy of DPA Picture Alliance Archive/Alamy)

19. Manning a GPMG on the *Atlantic Osprey*. (Author's collection)

20. Having operated from 1955 until 2005, the United Kingdom Atomic Energy Authority Constabulary was replaced by the Civil Nuclear Constabulary. (Crown Copyright/National Archives)

# 10

It is impossible to just walk up to a reactor building and head inside as if it was part of a factory or industrial site. When constructing a nuclear power station, one of the first things that will happen is that the technicians and project directors designing the site will locate the intended space for the reactor core on the best land that they can (thinking about many different and even improbable variables), then build the supporting infrastructure around it. So whereas the reactor is the heart of the station, it may not physically be in the middle. It will also be built on foundations that are designed to withstand the heat of any reaction that goes critical (a self-sustaining nuclear process – one that makes power), and around it would be built shielding and protection to contain the worst of the effects should a meltdown occur.

Engineering and scientific processes aside, the reactor is also designed to prevent any unauthorised access, with guard positions, checkpoints and high-tech systems limiting or preventing movement by unauthorised persons. They also prevent people from leaving should there be contamination of the individual or radiation detected inside the building. Being such a secure structure, with countless cameras and guards plus the health physics team on patrol, there is no need, unless directed by the regulators, for the

CNC to frequently access the reactor site. However, every officer will have undertaken the training necessary to allow them to enter the most vital parts of the building and would also have become familiar with the layout by joining the guard force on their patrols as they secure the facility.

The reactor at Oldbury-on-Severn, now decommissioned, is a labyrinth of corridors, access hatches, ladders and stairways. It would be easy to say that it looks like something out of a science fiction movie, but in truth sci-fi is more than likely based on this kind of engineering. As with any reactor, it requires the constant presence of scientists, engineers and support staff to monitor everything, even while offline. Under international agreements and subsequent Acts of Parliament, certain conditions have to be met by the site owner in order for them to be granted a licence to generate power through the process of nuclear fusion. For this reason, CNC officers sometimes refer to power stations as 'licensed nuclear sites' – those that generate power or store fissile material (nuclear fuel) – and 'non-licensed sites', which either store low-level waste (tools or uniforms contaminated during engineering tasks) or have entered the decommissioning process.

By the 2000s, Oldbury had been generating power for forty years and was undergoing an inspection to ascertain whether the reactor's life could be extended. Despite the best efforts of those who built the power station, eventually the infrastructure and reactor would one day need to shut down for good, and the century-long cleanup process would begin. The reason for such a long decommissioning is because of the amount and types of radiation that are present, particularly at the reactor site and cooling systems. However, not all radiation is immediately harmful to humans; some actually occurs naturally all over the world. For example, alpha radiation will not penetrate skin but could be harmful if ingested. Beta radiation, meanwhile, will easily harm living tissue but can be stopped by shielding and distance, whereas gamma radiation contains a lot of energy and can pass through even the most substantial barriers. As we were constantly reminded by the industry regulators, time, distance and shielding were the best defences against these high-energy particles.

The effects of radiation on the human body have been well documented. Because of conflicts and the developing nuclear industries around the world, there have been many recorded instances of ionising (high-energy) radiation coming into contact with living tissue. There is sufficient energy in some radioactive material to alter the structure of the cells in the body; no wonder security measures such as those provided by the CNC are employed. Nuclear proliferation, as it is known, is the acquisition of radioactive material by rogue states or terrorist groups who have the resources to construct either nuclear weapons or a 'dirty bomb'.[1] There have been instances of certain terrorist groups that have stated their intentions to build this kind of device. International agreements, overseen by the International Atomic Energy Agency (IAEA), seek to promote the peaceful use of nuclear energy and prevent its use for any military purpose. But agreements and handshakes are not enough to prevent nuclear proliferation and armed, physical protection that meets agreed thresholds is a global security concern. In Britain, the UKAEA works with the IAEA as well as other nations in securing nuclear fuel and radioactive material while also trying to gain a better understanding of the many benefits and pitfalls of the nuclear industry.

Radiation has the effect of easily damaging the DNA of cells, preventing them from regenerating correctly or even at all. This makes it a particularly effective but indiscriminate weapon when used as a bomb. At the end of the Second World War, atomic strikes

1 A dirty bomb is a conventional explosive device that is designed to expel radioactive material, as opposed to causing a nuclear explosion.

on Hiroshima and Nagasaki caused countless deaths not only from the infernal explosive blast but also from the radioactive aftereffects of the bombs. It was impossible to clear the cities of such large numbers of wounded civilians, and with the roads and railways damaged beyond repair it was difficult for aid to arrive in sufficient quantities to help those caught in the strike. The water and food that remained in the city soon became contaminated and caused further casualties from radiation poisoning and dehydration. In Hiroshima alone it is estimated 120,000 were killed, but the true number will probably never be known.

The bombs had far-reaching consequences and began a global arms race, defining much of the subsequent Cold War era. This also led to an expansion of civil nuclear projects, which could be just as dangerous. On 10 October 1957, there was a nuclear incident at Windscale in the UK; the site would go on to be known as Sellafield. A fire raged for three days and produced a significant amount of radioactive fallout, but this was minuscule compared to the accident at the Ukrainian power plant in Chernobyl in 1986. The well-documented Chernobyl meltdown has been the subject of documentaries and dramas for decades, partly because of the secretive nature of the Soviet regime but also because of the sheer impact of the incident, which had a global reach.

On a smaller scale, poorly understood practices and safeguards can likewise be extremely hazardous. In September 1999, a Japanese nuclear technician was exposed to a flash of gamma radiation that lasted no longer than one-third of a second. However, sufficient radiation was released in that time to irreparably damage Hisashi Ouchi's body. Over the course of the next few months, every part of his body began to fail. As his cells were unable to regenerate, doctors at the University of Tokyo Hospital reported that he was literally melting both inside and out, with his skin sliding off his body when not bandaged in place. He died after eighty-three days, leaving a wife and daughter behind. In November 2006, meanwhile, Russian defector Alexander Litvinenko was poisoned in London with an alpha type of radiation known as polonium-210 in what was widely regarded as an assassination. Similar to Ouchi, within three weeks he succumbed to the effects of the radiation and died. He was buried in a lead-lined casket.

Despite the well-understood effects of radiation, most CNC officers were not issued much in the way of protection other than a military-style respirator. I imagine that if an accident did occur then nothing short of a spacesuit would be of much use, and that would make policing very difficult. However, radiation was not the only hazard on site. Various gases, chemicals and explosive materials were a constant presence – and these needed to be secured in a way similar to the most vital areas.

As firearms officers, much of our training focussed on the 'backdrop', the area behind where we were shooting. You had to know what might be hit if you were to miss a shot or over-penetrate. In a shopping centre, you might hit an innocent bystander. Near a road, you might hit a vehicle and cause it to lose control; in a nuclear power station, you might hit a pipeline full of sulphuric acid under high pressure, which would melt the surrounding area as well as anyone unlucky enough to be nearby. This applies equally to where officers would take cover when dealing with an incident. A substantial piece of furniture between you and the threat can literally be a lifesaver, but the CNC must have knowledge of the whereabouts of various chemicals, containers and stores. In order to maintain this level of understanding and to be aware of changes, internal patrols on site and occasionally in the more vital areas of the power station are essential.

On 30 May 2007, I was sitting with Cal in the ARV on the crossroads that led to the site. He had just returned from a response driving course hosted by North Wales Police, and I was due to attend the same course the following month. Mack was away on a CBRN (Chemical, Biological, Radiological and Nuclear) course, training to become a CBRN reconnaissance team leader along with another officer from Oldbury called Andy. Andy was a svelte officer in his late twenties who seemed to live off energy drinks but never seemed particularly awake. Perhaps he had built up a caffeine tolerance over his years with the police. With Mack away training, Aled was acting as the shift sergeant for our team for the next few weeks. He was on site with the security team, tying up some loose ends and managing the admin that was supposed to be the governor's job.

As we maintained our watch over the roads leading to and from the site, I quizzed Cal about his experiences and what kind of things I might be expected to do on the response driving course. So far I had only read a couple of books on police 'roadcraft' and held no driving qualifications except for my normal driving licence. It sounded great. Three weeks away with a different force, blue-light runs, skid training and night driving in the North Wales countryside.

'You'll love it, Matoo.' Cal had for some reason decided that from now on I was not Matt, or Matty (as Dave called me), but Matoo. I didn't mind though. Cal was so easy to get along with that he could have called me anything and I would have felt his natural sense of humour come through in even the harshest of nicknames. 'Great accommodation at a decent hotel, loads of grub and no shifts for three weeks!' That sounded good to me. 'But it is a steep learning curve,' he went on. 'Just remember. Position. Speed. Gear. Acceleration. Everything happens in that order. If you can build that in with a running commentary of your drive, then you'll find it no problem at all.' I had read about position, speed, gear and acceleration before, in the roadcraft book that I had been sent by the CTC, but took his advice on board all the same.

As we were talking we heard a low rumble, like a truck had just driven behind us.

'What was that?' asked Cal, looking out the rear window and expecting to see a tractor in a ditch or a light aircraft hanging out of a tree.

'No idea. Maybe one of the farms has had an accident. Tractor crashed into a shipping container maybe?'

Cal looked back with his finger pressed against the glass of the ARV window. 'I think that might have something to do with it.' He was pointing in the direction of Oldbury and more specifically at a black cloud that was starting to rise out of the power station.

I didn't know what to say. I didn't know what to do. I did, however, double check that the ARV windows were rolled up and the air vents shut. In all the training we had done, nobody had told us what to do in the event of something going critically wrong with the site equipment. They had, of course, joked about 'kissing your arse goodbye', but industrial incidents were not a police job. That was a job for the scientists and engineers, who were paid the big bucks to make sure things like the Windscale Fire never happened again.

'So what now?' I asked.

As if to answer that question, the ARV phone started to ring; it was Aled.

'Yeah? It's Cal,' my crewmate said as he answered the call.

I could hear Aled talking to Cal through the speaker on the handset. 'Just to let you know,' began Aled, 'something on site has blown up.' His slow, rhythmic voice didn't betray any signs of concern or worry.

Cal looked round to me while taking out a notebook and pen just in case Aled had some specific instructions for us to follow.

'It sounded like it came from the turbine hall, and the site fire brigade are on their way now.' That sounded like good news and bad news all rolled into one: good that it was the turbine hall and not part of the reactor site; bad that the fire brigade were required. I had visions of most nuclear incidents starting small and growing out of control – one of the perils of playing with atoms.

'The Avon Fire and Rescue response is on the way, as well as every appliance[2] that they can muster–'

Cal cut him off. 'So what do you need us to do?'

'Nothing you can do really, whilst you're on external, and I don't want you back in case things get worse. The two on internal patrol are at the gates now, keeping overwatch on access control. You two make sure that there is no traffic build-up and that the appliances can get in. Where are you right now?'

'We're on the cross.'

'Okay, good, you stay there and stop any traffic heading towards the station – make sure those appliances are not obstructed.'

'Got it.'

Cal looked back to me. 'You catch all that?'

I nodded. 'I guess we're okay here. It's the last junction before site… I wonder what those firefighters are thinking?' I didn't know which ones I meant as I said that. All of them, I suppose. The power station fire crew were all volunteers from the various departments. They mustered for fire drills and industrial accidents in the same way that retained or volunteer firefighters might anywhere in the world. They got a little extra money for doing it too, but probably never expected an explosion during their careers. The full-time crews, coming from the nearby towns and the city of Bristol, were probably feeling a bit nervous about a fire at a nuclear site as well. I was sure they were having thoughts about Chernobyl. And then what about our police contingent? We were less than a kilometre away, and the sarge and internal patrol team were on site as well.

---

2 Fire engines are often called 'appliances'.

I wanted to check which way the wind was blowing, but I was reluctant to get out of the ARV or open the window so I looked to see which way the trees were swaying. We heard Oscar Yankee Two call in to Harwell. The channel was for policing purposes and connected us to the entire BCU, meaning that every CNC officer in the southern unit was up to date with what was going on. After a brief appraisal of the situation, they informed the duty inspector that they would report any further updates by phone as the messages were likely to be very long and block the airways.

'I can hear them coming,' I told Cal as the familiar two-tone siren of the fire brigade came within earshot.

The closest crew was in Thornbury. The station was staffed during the day but was in the hands of volunteers at night. Luckily, the crew was available and not on a different incident somewhere in the area. In less than five minutes they were driving past us and down the road to the fire. Looking through the windows of the truck as they went by, I could see that the crew were still fastening their breathing equipment; that, or double- and triple-checking that they had airtight seals. Within minutes, another appliance came by. Then another. And another. Then more. In total, ten fire engines came by in less than twenty minutes. The site had two fire appliances of its own as well – one a typical fire truck and the other a Land Rover designed as an 'incident support vehicle' – so there were twelve emergency crews – more than sixty responders.

'Well, hopefully it's sorted. Do you think I should give Aled a call? Get an update?' asked Cal.

'Probably best not. I guess he's talking to the site or fire brigade anyway. He'll let us know when we're needed.'

Cal nodded in agreement then angled his seat back a little bit. 'Well, then I'll just make myself a bit more comfortable and wait for a call.' He put the mobile phone on the dashboard and turned on the radio. The ARV still had a media player built into the dash, and Cal enjoyed the classic 1980s and 1990s hits played by the local DJ.

*Fair enough*, I thought. I could see this being a long and very static shift.

It turned out that part of the non-nuclear-generating machinery had overheated that day. This 'step-down' transformer was huge – easily bigger than a three-bedroom home – and designed to 'transform' the immensely powerful voltages produced at the station, reducing them (stepping down) for use on the power grid. Seeing the charred hulk as we drove back into the site was a reminder that even the most careful of site operators had to be vigilant to the point of paranoia.

It was speculated at that time that perhaps the internal machinery was worn out, or there was a power surge that somehow went undetected. In truth, there could have been any number of reasons, but these were never made public. What *was* made public, though, was the incident itself, and the successful prevention of an even larger problem at Oldbury-on-Severn. The fire was extinguished by the volunteer crews and fire suppression system on site, before the regular firefighters had even arrived. As a security precaution, the entire site was shut down and the staff were mustered to their evacuation points. British Nuclear Group reported the incident through the Reuters news agency later that day, but barely anything was mentioned in the local or national press.

Finishing our shift, Aled filled us in on some more of the details that he had gleaned from the emergency engineers on site. No doubt further details would be added over the coming weeks and months. We knew, however, that a significant amount of checking and some repair work would be required before the reactors were brought back online. In no time, the regulators would be descending on the site in droves to find out the root cause of the problem and to decide whether Oldbury would be permitted to go on generating. It was also likely that the local nuclear protest group, and larger members of the anti-nuclear lobby, would catch wind of the 'disaster' or 'incident' at Oldbury. This could lead to physical protests at any of the sites where the CNC was deployed, or it could lead to holdups in the construction of the new Hinkley Point C reactors that had just been delineated, ready for the first parts of the massive project to begin.

Considerably bigger than Oldbury-on-Severn power station, Hinkley Point (known as Hotel Papa) was split into two separate areas. The 'A' site was decommissioned and undergoing care and maintenance while simultaneously being cleaned up. The 'B' site was generating, and, like Oldbury, it used the water of the Severn Estuary to cool its processes and to generate steam for the turbines. At that time the 'C' site existed only on paper and construction had not yet begun, although as of 2025 it is one of the biggest engineering projects in Europe. As it was not too far from our power station it was not uncommon for officers to be sent to Hinkley Point on an overtime shift – and one day it was my turn.

I arrived at the security gatehouse, which doubled as a police station, on a weekday evening. It was short-notice overtime, owing to there not being enough officers to cover the shift that night. A sergeant who had previously been a PC at Oldbury was running the shift and was joined by two brand-new probationers, each of them with only a few weeks of experience. As I had just over a year of experience on them, the sarge paired up with one officer while I paired up with Becca, a tall, red-haired girl from the north of England. I already knew that he would not pair up with me owing to a disagreement between us while he was still at Oldbury as a PC. Even so, it was better to have someone with experience accompanying each of the new officers. Save for the unfamiliar surroundings (I had only been to Hotel Papa a handful of times), the shift was shaping up to be one like any other. Outlining the shift strategy, the sergeant decided that he would be first out on patrol with his crewmate. That meant that Becca and I would start with an internal patrol, walking the power station.

It was a warm summer night, so patrolling on foot around the power station felt quite pleasant. Even though most of the site personnel had retired for the day, the odd engineer or security guard could still be spotted attending to their tasks. It felt quieter than Oldbury, probably owing to its larger size. I didn't know my way around the area but the shape of the site meant you always ended up where you started. It felt a bit like walking around an unfamiliar town centre; similar to what you are used to, but also different.

'Have you been in long?' Becca asked me.

'I'm just about to go on stage five, so only a year. I've been to Hinkley a couple of times, as well, but I suppose you will know more about it than me.'

She shook her head. 'I don't know anything yet. The sarge just said to follow your lead, so that's what I'm going to do.'

It was good to be left alone to be able to get on with the job without too much interference. It was also good to be trusted to 'puppy walk' a new officer – much like I had been when I started my operational career at Oldbury. I thought it was odd that her instructions were just to 'follow my lead', however. I was half-expecting to be ordered to stay on site for the full shift, not just because of my own inexperience but also due to the bad blood between me and the shift sergeant.

The sun had set by the time we started our external patrol. Becca was driving as she knew the roads better than me. She was living locally, having moved to the area a few weeks before her first shift.

'Full moon tonight. Better keep an eye out for the Witch Watch.' I said as we picked up speed, heading away from site.

'The what? The Witch Watch?'

'Yeah, the Witch Watch. They're harmless enough but they come out when the moon is full to protest nuclear power. We've seen them at Oldbury a few times but they're more local to this area. You'll bump into them before long. They don't cause any trouble though, except for alarming the local residents and security guards… Just by being there and being dressed in their garb.'

The Witch Watch based their nuclear protests on the lunar cycle, which is twenty-nine days long. Calling it a 'moonthly' vigil, they would hold gatherings in public spaces near the power stations. Threatened with a new nuclear power station (Hinkley Point C), they had become active and were reported to the officers on patrol by CNC special branch, who kept an eye on the local and national protest groups.

Honestly, I did not have a problem with these people, dressed in black and reciting their homemade incantations; I just saw it as a group of protesters with a strong common interest. Unlike certain other groups, they did not block traffic or try to undertake any direct action – at least in a sense that I could comprehend. In fact, before I had deployed to Iraq, a 'witch', who I did not know was a witch, had presented me with an 'anti-terrorist curse' that she and her husband had written. I did not know them particularly well, but as I came from a small village my impending deployment was common knowledge to my neighbours. I was touched that they took the time to craft some words meant to keep me safe, and I still have the handwritten enchantment to this day, along with other important or meaningful items I have collected over the years. I could not say whether the spell worked, though.

# II

Dave was right when he said that the stage five review course would be easy; it was nothing more than a formality. However, while there I was surprised to discover that many of Course Five had already left the Civil Nuclear Constabulary. Rachel had left following a disagreement with the firearms training department; she and another officer claimed that the CNC had discriminated against their petite sizes by not issuing them with guns with smaller pistol grips. It even made national news, which might have been the CNC's first appearance in the press since the announcement of its creation back in 2005.

Several officers from Sellafield had decided that the job was not for them – a lot of shift work and static patrol duties. Those could be difficult for anyone to regularly undertake, but the CTC had been up-front and honest about police duties, especially at Dounreay, Sellafield and Harwell, where this kind of tasking was most common. Two officers from Harwell had already applied to join and started their training with Bedfordshire Police, and Ross, posted to Dungeness, had joined that power station's guard force (who were on substantially better pay and conditions than the police).

In all, 50 per cent of the course had moved on within the first year of their policing careers. Sergeant Buckden, who was running the stage five review course, said, 'It happens every time. I can tell just by looking who's going to stay and who's going to jack it in.' Regardless of the attrition, it was good to catch up with those who were still in the job over a pint in the Moon Bar.

The officers who had previously made up our recruits' course were happily swapping war stories and talking about incidents they had been involved with or arrests they had made. Banksy had chased a trespasser who tried to run through the gate at Harwell.

'The nugget just ran through but I dannae know why he thought he could get away with it,' he noted, given that Harwell is a labyrinth of fences and buildings with a complex network of roads that even the site staff have difficulty remembering.

Banksy was severely admonished for leaving his post, though – the gate was unmanned during his foot chase. He was reminded that there are mobile patrols and dog handlers to deal with runners.

'It only took a minute to catch him,' he added. 'The inspector was probably looking for some fuel for his CV. Someone to offer "advice and guidance" to.'

Mark, who was now at Dungeness, had found a foot while on patrol – a human foot. It had washed up on the shore of the beach beside the power station. Even though it was probably from a burial service at sea, he and his partner had to remain with the appendage until a coroner's officer arrived to take it away. And Knoxy at Sizewell mentioned that one of our own officers was arrested and under investigation for being in possession of an illegal substance. Apparently her neighbours had reported the distinct odour coming from the flat that she shared with her partner. Knowing full well she was a police officer, her neighbours raised their concerns with Essex Police.

'She was always cagey about her life away from work. Never talked about family and friends and hobbies. She was so intense… I guess it helped her to unwind…?' Paul was thinking out loud. 'Well, I doubt we'll see her again. That's two we've lost to professional standards at Sizewell. Did anyone ever meet Julian? He turned up stinking of booze. The skipper noticed before he got tooled up, and he ended up getting nicked for driving under the influence. But the locals let him rest up for a while so by the time he was tested for booze in the blood he was all but sober. And then, then he got transferred over to your lot at Harwell.' He turned to Banksy.

'Aye, we know all about Jules,' he replied. 'Bullshitter. I don't know what's with that nugget, but I don't trust him. I'm glad the boy's on a different shift to me.'

When it was my turn to contribute to the conversation about our first year of experiences, I decided to go with my most recent encounter. The week before coming to the course I had been on patrol with Mack, on a night shift and heading into the nearby town.

'On the other side of the road the sarge noticed a Transit coming the other way with cans all over the dash,' I explained. 'The driver obviously saw us and put his foot down, making off in the opposite direction and turning into a residential area. We quickly turned around and got the lights on, and followed him onto the street.'

The CNC does not have a pursuit policy, outside of the escort group, but is able to follow and observe where necessary.

'Anyway, Mack came from Gwent Police and I guess his policing instincts kicked in. The street had lefts and rights and junctions all the way down its length but in no time we pulled up alongside a badly parked van. The guy inside had dropped all the cans and things out of the passenger window and was shuffling back into the driver's seat ready to make off again when Mack parked in front of him and got out. I let the locals know what we had and went to join him. I was surprised the driver was able to find the ignition and start the van. He could barely speak so Mack arrested him there and then put him in the back of the wagon while we waited for the locals.'

There were nods of understanding from the group, as this event was not too dissimilar to events that most of us had dealt with by now.

'Mack said to me while we were waiting, "I bet you never did any real police work until I got here!" and slapped me on the back, taking the piss out of me.'

He was probably right. Within the first year I could probably count my 'real police' events on one hand.

When I returned from the course, I found that the TRG national firearms training instructors (NFIs) had come to Oldbury to train the officers on the use of Taser – a conducted-electricity device that looks like a pistol and is designed to incapacitate rather than kill. At that time, Taser was a new weapon system that had only been approved by the HOSDB[1] in 2003. The slow roll-out across firearms officers in the UK had finally reached the CNC executives, who agreed that OPU officers should begin carrying them on duty. Over the course of two days we were to be trained in the use of the device by Dave and Geoff, two venerable members of the UKAEAC who were now NFIs on the Tactical Response Group.

In truth, Taser was very simple to use. It had a laser to assist aiming and could only fire 21 feet, as opposed to the hundreds of metres some of our other weapons could manage. It didn't have any recoil, making it very user friendly. It held a cartridge which, when discharged, released two probes connected to the weapon's power source. These probes shock the target, causing NMI (neuromuscular incapacitation), turning their major muscle

1 The Home Office Scientific and Development Branch is the government department responsible for the approval of all police equipment used in the UK.

groups rigid and rendering them unable to move. It could be tricky to reload, though, especially when under the stress of an incident or scenario-based training. Once discharged, the device needs to be deactivated before a new cartridge is put in. If an operator failed to deactivate before reloading, there is a very real chance they themselves would be shocked.

Our previous less-lethal options had been CS spray and the baton gun. CS spray is not actually a gas but rather a solution with tiny crystals that irritate the mucous membranes of the mouth, throat, eyes and nose. It can linger on clothes or in the air (especially inside buildings), so there is always the risk of catching a lungful of the irritant if it is not used carefully. When it is deployed, the officer is supposed to shout 'GAS!' before using it, to warn others of the hazards. Shouting 'gas' at a power station would cause a huge panic, however, so the CNC's specific training took into account the danger of miscommunication as well as the risk of the many air ducts on a site swallowing up CS and depositing it into sensitive areas.

The L104 baton gun, meanwhile, was sometimes called 'the blooper' by those who carried it because of the distinctive noise it made when fired. It sounded like a deeper version of a champagne cork being fired. My opinion was that a cork might well be a more useful projectile, as the 37mm AEP (attenuated energy projectile) never came across as a particularly effective round, at least to me. Years earlier we had used 37mm rubber bullets in the army, but the AEP was noticeably softer than a rubber bullet and made a good chew toy for the canines in our dog sections. On the other hand it was a fearsome-looking gun, with a wide barrel and extendable stock; it partly resembled a grenade launcher. The addition of Taser to the CNC arsenal was effective and very welcome.

Taser was not the only significant change to the CNC at this time. A few months after the new equipment had been deployed to the units, Chief Constable Bill Pryke had left and been replaced by Chief Constable Richard Thompson. The 'chief bobby' usually only does five years before moving on, so Mr Pryke retiring was not unusual, however his replacement, Mr Thompson, was a unique selection by the CNPA (Civil Nuclear Policing Authority). Not only was he not a police officer, but the rumours were that he was from a special forces and intelligence background. The CNPA and the government clearly had some intentions regarding the future of the CNC, and by bringing in a former spook they were indicating that things were going to change.

While I was on patrol with Gemma, our sole female officer at Oldbury now that Nix's secondment had ended, Mr Thompson visited as part of a tour of all his stations. Being a very busy man, we were only able to speak to him briefly before resuming our duties. Handsome and olive-skinned, Mr Thompson looked like a rugby player. I got the feeling that he deliberately wore shirts that were one size too small for him as well. He was shorter than me, but powerfully built and with a penetrating gaze. Gemma let me know her first impressions once we back inside the wire, walking our beat.

The topic of conversation soon changed to Gemma's other burning desire: to transfer to Dungeness and be closer to where she grew up in Kent. By now, Mark had moved to that unit to be within easy reach of London. It seemed to me that a lot of officers at Oscar Yankee were only with us temporarily. For my part I was more than happy, but I could understand their situations. I was in a part of the world that was familiar to me, while many of them were a long way from home.

Gemma and I were due to attend a qualification shoot over the next few days, so she had been taken off night shifts and put onto the day shift prior to the training. Every quarter we were supposed to take a requalification test, which typically lasted all day. The following day we would normally practise some scenarios and firearms drills. To achieve this, our instructors from TRG would make use of another constabulary's range and tactics facilities. In our case we were off to West Mercia, near to where Dave lived, to use their HQ training area.

A requalification shoot is almost exactly the same as the initial test undertaken by recruits. Shooting from various positions and distances, the officer is assessed on all the core criteria for a police firearms officer. If they fail, they can receive additional coaching and reattempt the test; however, they would not be able to deploy on patrol and would be stuck doing unarmed duties until they were qualified again. If they failed a second time, they would be sent to Sellafield for some further instruction to get them back up to standard. This was very rare, but officers did occasionally have 'off days' where they missed a pistol shot from 20 metres.

As we patrolled around Shepperdine and the area east of the power station, Gemma asked me a question.

'How come you don't speak like Cliff and the other locals round here?'

She had a regional accent herself, being from Kent.

'I don't know... I never really think about it.'

'Well you sound posh,' she replied. I wasn't sure if that was a compliment or an insult, but either way I wasn't too bothered about it. 'Do you come from a posh family or something?'

We carried on our patrol, Gemma in the passenger seat and me driving while undergoing this unexpected interview.

'And Mack says you're high strung,' she concluded.

I knew it wasn't worth discussing in detail, but this had aroused my curiosity.

'What do you mean, high strung?' I asked.

'Oh you know, you overthink things and ask lots of questions. Things like that.'

What I had taken for enthusiasm was possibly not coming off in the way I had hoped.

'Well, thanks for that Gemma. I'll try to restring myself at some point.'

We crossed over the main road leading back to the power station to patrol the other half of the beat. As I drove, Gemma got her phone out and started to text her boyfriend, smiling as the messages came and went. I couldn't help but feel as though her attention wasn't entirely on the patrol. I thought to myself about some of the practical jokes that other officers had played on distracted or sleepy officers. There was often manure on the roadside that you could park next to, so when you switched seats the passenger would step into it as they disembarked. But then I would have to share the odour, and I didn't want that. And besides, what some people see as playful banter would be considered bullying by others, and remembered for a long time.

'Let's park up and do some statics,' I told her as she tapped away on her phone. 'After that, you can drive.'

I was enjoying a cup of tea after patrol and had just settled down to go through some paperwork when the station telephone rang. Beegee, the unit deputy, picked up and began speaking to the person on the other end. The office only had thin walls, so it was impossible not to catch what someone was saying – especially if they had a broken volume switch, which seemed to be the case for Beegee.

'Oldbury OPU, station deputy here,' he answered, probably while reviewing the vast amounts of admin that seemed to consume his day.

'You're gonna need to say that again mate, and slower.'

That caught my attention.

'When? Right and, errr, are they on site?'

I decided to poke my head around the corner of the office to make it easier to hear. Looking up from the desk, phone braced against his shoulder, Beegee pointed at me and indicated to stay put. I did as I was told while straining to hear the voice on the other end of the line.

He hung up. 'Right, Matt, where's Gemma?'

She was on her break as well, so I gestured in the general direction of the kitchen. 'Making her lunch I think. What's going on?'

Beegee got up from his desk and made his way to his equipment locker. 'Tell Gemma to stay here and run the office. You come with me on to the site.'

This all sounded mysterious and out of the ordinary. The site security team rarely asked for a police presence, but I guessed that was who had been on the line.

'Sure, I'll tell her. What are we supposed to do on site?'

Walking the brief distance from our station to the guardhouse on site, Beegee filled me in on the details.

'So a couple of contractors have taken a load of the scrap metal from that project up behind the boronic acid store. You know the one?'

I did. The area was part of a small complex used to store 'dirty' equipment – as in the tools and equipment that had been used in the reactor areas. Once the tools had been used once they were never used again owing to the risk of radioactive contamination. The building was being reinforced with new walls and ceilings, meaning that some of the old parts of the structure had been stripped out. The waste material was not thrown into a skip or lorry as might happen at a factory or leisure centre; but was placed carefully into a fenced-off area, awaiting disposal.

'Security saw a couple of the scaffolding contractors loading the scrap onto their truck. Mr Surl wants us to look into it.'

On the face of it, this was potentially not a serious matter. I supposed the contractors had assumed the metal was waste and that they would be doing the site a favour by disposing of it, while probably making a few quid themselves. The problem, however, was the possibility of contamination. If contaminated metal was recycled with standard scrap, there was a good chance that people would soon be drinking from radioactive coke cans or buying irradiated goods from the local hardware store. After all, there is no visible difference between scrap metal and radioactive scrap metal.

'Ask Gemma to recover any CCTV,' Beegee said. 'We'll talk to the guards and contractors... Get this figured out.'

The contractor's truck was loaded with scaffolding poles, wooden boards and sheets of steel that had once made up the floor of one of the buildings. Mr Surl, the site security manager, was sitting in a liveried security vehicle, watching from a safe distance while his men erected a fence around the flatbed. The two contractors stood off to one side with a guard, awaiting the arrival of health physics. I had just heard over the local VHF radio that a physicist would be on the way, Geiger counter in hand, just as soon as he was in his radiation suit.

'Health physics is on the way sarge,' I said, as I reached into my belt pouch for my notebook and pen. I knew that Beegee would be doing the talking and that I would be recording the relevant details.

Ignoring Mr Surl and heading to the guard with the contractors, Beegee took off his cap and began the investigation.

'Alright mate, we have to look into this,' Beegee said while addressing the older of the two scaffolders. Their flatbed indicated that they had come from Birmingham, so I wrote that down along with the company name.

'Give me your site passes, gents.' Beegee extended his hand to the contractors but the guard handed them over instead, having already taken them. Beegee looked at them and checked the photographs against the faces of the two men. I took the passes from him and wrote the details down, then handed them back to the guard. Mr A and Mr Y probably didn't realise the gravity of their actions. Mr Surl, on the other hand, very much did. Outside the wire, this kind of incident might well be dealt with by way of some carefully worded advice, or perhaps a summons to court. But there were additional factors in play on a licensed nuclear site and there were security standards that were enforced by a dedicated regulatory agency. This had to be investigated.

Having taken their details, Beegee introduced us to the two men and went on to explain that the site security team had requested a police presence. As he spoke I wrote down their responses. The sarge had cautioned them and explained that he suspected them of theft.

'It's not theft though, is it?!' protested Mr A. 'I know the law. If it's abandoned then it's got no owner. I became the new owner by taking it "legally" for myself.'

I could see where he was coming from, but it wasn't much of a defence. The fact of the matter was, Mr A and Mr Y had opened a gate, entered a controlled area and taken the scrap without seeking any approval. By interfering with site structures, they were now suspected of criminal as well as industry-specific offences. With that in mind, I stepped away to call Gemma on the radio.

'Oscar Yankee Seven Four Four, from One Four Two.'

Gemma was waiting in the police station as Beegee had instructed. 'Go ahead.' She sounded very different on the radio than she did in real life. I could barely hear her south-eastern accent.

'We're with the security team now. We need you to head over to the gate house and recover some CCTV footage.'

She didn't need much more instruction than that. Gemma was more than capable of figuring out what footage, where from and what timeframe. The guards were probably already putting together a copy for their own records, so securing a second copy should not be too much of an effort. I thought it would also be a good idea to get her to come on site as well; that way, once this incident was over, we could conduct an internal patrol while Beegee submitted the GPR. The regulators would still expect a full service of police duties to be carried out, even in the event of a breach of site regulations.

'Can you come onto site once you're done?' I added. 'We should go on patrol after this.'

Her short response came instantly. 'All received.'

Rejoining the scaffolders and sergeant, along with the security guard, I saw that Beegee had his pocketbook out and was making notes. He side-eyed me as I returned, probably annoyed that he had to start making notes when he preferred someone else to record the details; he preferred to only put pen to paper once there was a cup of coffee to hand. Mr Surl was still watching from inside his car. I guessed that he was making some notes or had a reason to be there; otherwise he was just a spectator, which would seem quite strange. Then again, he was quite intense at the best of times – at least that was my opinion from the few times I had met him.

Beegee was questioning the guard now. I didn't know the guard's name but I recognised him. He had long hair in a ponytail and was either from Australia or New Zealand as far as I could tell from his accent. He was tall, so easy to spot when he was conducting his patrols. The guards each carried a GPS-enabled 'baton' that had to register their presence at certain parts of the site within a given timeframe. These areas were sometimes quite inaccessible but the procedure ensured that the guards covered every inch of the site during their shifts. I had often seen this lanky, windswept guard making his away across the car park on a dark, wet night. The guard was answering Beegee.

'... Then this afternoon the gate had been opened and I saw this truck and these two fellas in there, helping themselves.'

I started writing and added my own question.

'What time in the afternoon?' The CCTV would give me the answer, but it was important to get his story as well.

'Just before you came. You know, about twenty minutes ago.'

I checked my watch. He meant around half-past two in the afternoon. I wrote it down: 'approx 1430.'

'Then I asked to see their paperwork, guessing that they had some or that it was something that I had missed during the morning meeting.'

During the investigation, a health physicist turned up his white disposable coveralls, thick rubber gloves and holding a rectangular box with a coiled lead attached to a cylinder. These bright yellow boxes were all too familiar by now and I knew that he would soon start poking the sensor of the Geiger counter into the piles of metal and at the two contractors. The clicking of the equipment is almost constant, with no discernible pattern. This is because background radiation is emitted by just about everything in existence. The sound becomes more constant and rapid once a source of radiation is detected. Once it started 'leaping', you knew that there was a significant source of energy nearby and it would usually cause the operator to quickly glance at the digital display to ensure that the source was not going to cause some health issues if he remained near it too long.

The clicking remained steady as he passed it over the metal; no different to the sound from when it was turned on. When the probe was pushed into the pile of metal on their flatbed, the pace of the clicking increased. Despite his protective gear, he had no helmet or mask so we could tell that he wasn't overly concerned about the scrap. Next he scanned the two contractors.

'Were you wearing gloves?' he asked. Wide-eyed from hearing the Geiger counter clicking, the younger man indicated that they had been and that the gloves were on the seat of their truck. Waddling over with his equipment, he undid the passenger door and leaned in.

'What does that mean?' asked the older contractor.

'It means you're nicked,' replied Beegee.

Presumably the contractor was asking about the clicking of the Geiger counter, but it was not a serious amount of radiation at all, equivalent to the luminous hands of a wristwatch. However, some people would regard any level of radiation as serious. In reality, the background radiation of everyday life is in many cases higher than in the non-nuclear areas of a power station, which is usually kept spotless. Radiation from the soil and from being at different altitudes or in different areas on the planet mean that the human body has built up a tolerance to 'normal' levels.

'We better sort this out in the site security office.'

By now Gemma had joined us, having made her way up from the gate to our location on the power station. This incident would now be handed over to CNC's Special Branch, along with the ONR security inspectors.

'Let me put them in a pod first,' added the physicist. These 'pods' were similar to airport security arches, but with sliding doors either side. Once shut, the pods, which were at the entrance and exit of the reactor buildings, were highly sensitive contamination detectors that could get more accurate measurements than handheld Geiger counters.

'Okay. I'll come with you. You and Gemma, take the long way back to the nick and do a patrol.'

As we patrolled 'the long way' back to the police station, I could hear Beegee on the radio speaking to the duty inspector at Harwell. Gemma was on a different channel, listening to the Avon and Somerset frequency, so I unplugged my earpiece, allowing her to listen to the transmission.

'I wonder what will happen now?' she asked to nobody in particular.

It was a good question. There was no way of knowing how the site or the regulators would want to handle this, as there were few significant incidents like this in the UK. Site security was necessarily very tight to prevent just this kind of incident, and clearly it had worked. However, I was sure that the ONR would be asking a lot of questions – as that was their habit for just about anything that came across their desks.

'I don't know,' I answered.

I did recall an event in Estonia where a similar incident had occurred. During the first week of training at CTC, our course sat down to a documentary about nuclear proliferation produced especially for the CNC. In 1994, three scrap metal merchants in a town called Tammiku broke into a compound and what they thought was an abandoned shed to recover and sell the scrap inside. They were local to the area, and as far as they knew the shed had been locked and left untouched for years, making it fair game for their business. In fact, this was a waste storage facility from the Soviet era. Some of the metal that they recovered was later found to be caesium-137, a byproduct of spent nuclear fuel. It was also the same kind of radioactive element that Chernobyl had released into the environment in 1986. The theft went undetected, but twelve days later one of the men died of radiation poisoning – along with his pet dog. The sudden death baffled the medics at the local hospital until one of the other men was admitted as a patient. He needed treatment for severe radiation burns on his hands from where he had been handling the scrap.

Today was orders of magnitude less severe. As the physicist had indicated, and as I could hear on the radio during Beegee's report, there was no indication of an 'unintentional radiological disturbance'.

Back on duty after a few days off, Dave and I were patrolling near the water intakes and making our way back to the entrance to the power station. There had been some severe weather over the past few days, and the river was producing some impressive waves that regularly reached up the steep riverbank. Dave had decided that he was going to take advantage of West Mercian Police's latest recruitment drive and had been in touch with their human resources team. He was hoping that, once he had gone through all his training, he would be fast-tracked into their firearms unit. He made a very persuasive argument. Firearms training, he reasoned, was not a cheap endeavour, and even those who are selected to begin training do not always complete the course. As a nationally qualified firearms officer, he would not need to undergo the full training package and would save them a lot of time and money. Although he had been accepted to start his police training (again), he had yet to hear back about his offer regarding his AFO status.

As we made our way past the helicopter landing site (HLS) close to the site entrance, we saw an old man waving at us from the vehicle barrier that separated the HLS from the road. We didn't usually get visitors, but it happened occasionally as the areas around the power station were quiet and picturesque.

Dave waved back. 'He looks a bit old to be going for a walk, but good for him. He's probably doing more exercise than most people his age,' he observed.

The man looked to be in his eighties, and he continued to wave as we got closer to the vehicle barrier. The penny finally dropped that this was not some pleasant old fellow who wanted to thank us for being such upstanding officers; he wanted our attention, and he wanted it urgently.

Barely able to breathe or talk, this resident of Oldbury village had lost power to his house during the storm that had passed over South Gloucestershire the night before. This meant that the oxygen machine he used to treat his COPD (chronic obstructive pulmonary disease) had not been working and he was down to the very last bars of pressure in his portable oxygen bottle. We rushed him into the entrance of the police station, and I went to collect his oxygen machine from his car that was parked nearby. I plugged it into the power socket in the foyer and handed him the mask. He slumped against the wall and held the mask to his face, not bothering with the green elastic head strap. I hoped that it would work. I had been unable to breathe once before, so I knew a little about the kind of feeling he might be experiencing. Prior to joining the police I had punctured a lung in a traffic accident. I could remember all too well the panic as my lung protested painfully and violently as the air entered.

Dave called Harwell and gave them an overview of events. He seemed remarkably calm considering the man looked like he was literally suffocating in front of us. I unhooked my G36 and set it down so that I could check the gauges on his air supply. During my three weeks of recovery in hospital for my punctured lung, I had spent many boring hours looking at the devices to which I was hooked up and asking the nurses any questions that popped into my head. I had used a similar oxygen bottle when moving around while waiting for my chest drain to be removed and for my ribs to heal. As far as I could tell there was positive pressure, and I could see his chest rising and falling. The compressor that refilled his bottle was whirring gently and there was a hiss from the mask, which he held in place. His eyes were closed but he had the strength to hold the lifesaving equipment in position. As Dave was providing the details to the inspector, he started to open his eyes. He pulled the elastic strap back over his head and adjusted himself so that he was more comfortable on the floor. He still looked pale, and very unwell, so I went to get him a chair from inside the police station and helped him into it.

His recovery took mere minutes, and it wasn't long before he was able to give us the full story. Even though his house and the village was suffering from a power outage, he assumed that the power station would at least have some power that he could use to refill his portable bottle and run his breathing device. He was right. The buildings in the power station ran directly from the electricity produced on site – it was self-sufficient by design. Now that he was feeling better, and had a supply to see him through until the outage was resolved, he wanted to make his way back home. Dave stopped him.

'I really think you should quickly visit the hospital,' he told him.

I knew where he was going with this.

'No, I'm fine thank you officer, I'll just be on my way,' he insisted.

Dave knew that if something *did* happen to him either today or in the very near future, a coroner would surely note that he had been in contact with the police immediately before. By calling an ambulance, Dave would be doing us all a favour. Persuaded to stay put with a cup of tea and a hobnob, the old gentleman waited with Dave for the ambulance while I fired a quick GPR over to Harwell.

While I was on the CNC computer system, something caught my eye. Escort Operations, a department that oversaw the armed protection of fuel as it was moved around the UK and abroad, was relocating from Sellafield to the CNC headquarters at Culham in Oxfordshire. This meant that there were two vacancies for operations planners – and it was a constable's role. I followed the link and read about what the role would entail: planning the operations, liaising with other agencies, acting as support for senior officers while a movement was on. And it all happened during the day. No night shifts. With Dave talking about moving to West Mercia, I thought to myself, *Well, I could stay here or I could try something new.*

I decided to try something new.

# 12

I was sitting in a coffee shop adjacent to the constabulary headquarters at Culham Science Centre in Oxfordshire. Having passed the paper sift, I was waiting for a 'competency-based interview' for the role of operations planner in Escort Operations. I didn't really have any experience in planning, so I was feeling a little nervous. The interview panel would include a police superintendent with a fearsome reputation and a notoriously blunt manner – William Watson, who was usually referred to as Wally. This senior officer was in charge of all specialist units of the CNC, such as Special Branch and Escort Operations, and had been in post for quite some time. Even though he looked rough and ready, with a shaved head, large frame and sizable gut hanging over his belt, he knew all the right people, asked all the right questions and always got things done.

Sipping on a coffee, I was trying to determine why it cost so much to mix hot water and beans when Amber, William Watson's personal secretary, came to walk me to the interview panel.

'Are you Matt?' she asked.

Guiding me through the HQ building to the interview room, Amber explained that the chief constable had decided he wanted Escort Ops to be relocated to the constabulary headquarters, and unless the operations team were willing to relocate then they would be reassigned to positions in other departments. Along with two operations planners, they were also interviewing for a sergeant and two civvie positions: a secretary and an accountant. There had already been a few interviews that day, so I had seen some of the officers come and go, but I didn't recognise any of them.

Taking a seat in front of the interview panel, the superintendent introduced himself along with Ed Stevens, who would be the inspector in charge of the day-to-day running of the department. Ed had transferred across from West Midlands Police and spoke with a Birmingham accent. Overweight and balding, his well-groomed moustache meant he came across more like a veteran detective than the second-in-command of the CNC's principal firearms unit. He had recently recovered from a bicycle accident and months of inactivity had caused a mild amount of weight gain, he explained. For that reason, he couldn't fit in his uniform, which is why he was the only member of the panel in a suit. The way he recounted his experience was quite disarming; no doubt it was a device to relax the candidates before the interview began. There was also a sergeant who supervised the Human Resources department, which was mostly staffed by civilian employees and not police officers. He was there to ensure a fair trial, apparently, and also to ask the all-important race and diversity questions that were mandatory in all applications and interviews.

After a very quick hour, Amber returned and led me back through the maze of corridors and back to the coffee shop. She asked me how it went and I told her that I was quite confident.

'I was able to answer everything, and throw some good examples in. To be honest, though, I hate interviews and tests and things like that, so I'm glad it's over!'

She smiled and held the door for me as I left the building. Apparently I was the last candidate for the day. I realised that it was getting late and I would have a two-hour drive to get back home. No doubt I would be replaying the interview in my mind and hoping for a good outcome, as it had become clear that this was something I really wanted to do. The more I thought about it, the more interested I became.

A few months earlier, when I was on a short refresher course at Sellafield, some new recruits to the Marine Escort Group (MEG) had been on the range, getting to grips with their automatic weapons. Speaking to them, I found that their course consisted of a dozen volunteers that had all passed an application-and-selection phase that began not long after Mr Thompson had taken over. They were looking to increase the amount of officers available for international work. I remembered the recruitment advert coming out and wishing that I was out of my probation so that I could apply as well.

The interview wasn't the first time I had heard about the work of Escort Ops. While at the CTC, the instructors who had been on escorts before would talk about their experiences during lessons, giving examples of CNC activities and the application of policies and laws. There was plenty of talk from among our IFC about whether an officer would want to go to sea for months at a time, or take part in a days-long road convoy. Sergeant Buckden had been part of an international shipment of fuel from the UK to Kobe in Japan but decided that one trip was enough for him. He liked the money but not the boredom of the Indian Ocean.

There were also two officers at Oldbury who had been on secondment to Escort Ops as plainclothes surveillance officers for a job in Workington Harbour. The two officers had taken a three-week surveillance course hosted by West Midlands Police before being sent up to Cumbria to lurk around the harbour and nearby roads, checking every car number plate they came across and recording the details of everyone they saw. The idea was that nobody from the local area would know the 'new' dockworkers so they would just leave them to get on with it. Once they were done, the REG descended on the harbour to secure it as the MV *Atlantic Osprey*, a nuclear fuel carrier, came in to dock and unload something bound for Sellafield. It was a unique policing role that gave the CNC a lot of its purpose.

I spoke to one of the officers who had been in Workington before I started my application. We were on shift one night while he was on overtime. I didn't work with him very often as his crew were usually on rest days while my section was on duty. He had transferred in from Avon and Somerset so did not have to go through a probationary period. About a month after arriving at Oldbury, he was away on his course, learning how to conduct surveillance in built-up areas. It struck me as a bit odd, because he was not very inconspicuous, being well over 6 foot tall with ginger hair. Then again, maybe that appearance would have fit well at Workington Harbour.

'My brother works for Severn Trent water, who provide water for the city we were training in,' he explained. 'So I borrowed a spare uniform of his and bought some traffic cones and a clipboard. After that I could stop outside any shop or building and look like I was inspecting some pipes or drainage.'

It was a good approach. Nobody ever interrupts a man with a clipboard. At Workington Harbour he had been paired up with another officer for the night shift and sent to the harbour.

'The problem was, it's such a small harbour. And two blokes in a car hanging around on an unlit road looks a bit suspicious. Luckily Cumbria knew we were there and dealt with the occasional call from the public, meaning we could get on with the job. In the end we decided that static observations in a Vauxhall Astra hire car would be too obvious, so instead we split up. The day shift had it easier, with people constantly coming and going.'

'So what did you do? Just watch people?' I asked. It seemed like that job could be carried out with CCTV or even regular cops.

I still liked the sound of it though – a nice change from internal patrols at a power station.

'Pretty much. We had a few specific tasks to achieve during the job, but nothing much. The chief came to debrief us afterwards and he said that he was hoping that the CNC would be able to do more of this kind of work in the future.' He shook his head. 'My feeling is that it was just the one gig, despite what the chief bobby says.'

'How did it go, Matoo?' asked Cal.

We were paired up for the day and watched the morning in-muster at the site gates. Similar to police officers at airports, we regularly conducted high-visibility 'static patrols' at vulnerable points on the site, most notably the entrance gates. Every staff member had a card and PIN that they needed to produce at a turnstile gate in order to get in, creating a queue every morning when the site staff arrived for work. It was probably one of the most boring tasks that we had, but we understood the importance of it. Occasionally we would be joined by a security guard as well, mostly there to help those who had forgotten their PIN or were otherwise unable to get through the gate.

Our role, meanwhile, was to provide armed security and identify anything suspicious or dangerous during this busy part of the day. Larger sites like Harwell had previously had interlopers trying to sneak through as part of a protest. Investigative journalists had tried to get through before, as well as suspicious spouses who wanted to see if their partner was really 'just friends' with that new work colleague that they had been hearing about. There was also the terrorist threat, which was our reason for existing as a constabulary. A sturdy armed presence was required at all times, so static patrolling was a necessary evil for us on the operational side of the CNC.

'Yeah, it went okay, I reckon,' I told Cal. 'Hopefully I'll hear one way or the other, soon.'

The car park was filling up with traffic and the workforce was coming in thick and fast.

'It'll be good to get out of doing statics, I know that much,' I added. 'What do you think for next? Internal patrol or external?'

Cal carried on watching the staff as they came in, throwing the occasional 'mornin'' to the passersby. 'External, I think,' came his reply. 'Have a cuppa first though, and then see what's out and about.'

We loaded up the ARV and made our way past the site boundary and began our external patrol. As it was a weekday, we weren't expecting much traffic now that the in-muster had ended. The roads were quiet. We headed right at the crossroads and into Oldbury village. Making our way through the village, we soon found ourselves in rural Gloucestershire, among the farms and orchards that made up the area. Our route took us into the nearby town of Thornbury and then back towards the power station.

'What's that?' Cal asked.

At the same time I began easing off the gas and applying the brakes. In the centre of the road was an upturned tricycle, the kind of low-riding carriage that is powered by hand and usually associated with handicapped cyclists. Lying next to it was a long-haired woman, her dark hair crossing the white lines in the middle of the road. We hadn't seen any cars coming in the opposite direction, let alone cars with visible signs of damage. Maybe a car going in the other direction had hit her?

I put on the blue lights and parked the ARV. Cal jumped out and made his way to the woman while I set about closing the road so that we could work on establishing what happened and what to do next. I knew that an ambulance would probably be needed, but they would want some details as well, so I made my way over to Cal to see what was going on. Lying face-up with her eyes rolled back and foaming at the mouth, it appeared that she was having a stroke or seizure of some sort and had fallen off her bike.

'Okay,' I thought, 'that should be enough to go with for the initial call to the medics.'

I had to let the team on site know what we were doing as well. Beegee was running our shift today, and he in turn notified the local cops about the obstruction on the road and the casualty.

Fortunately, the paramedics arrived quickly. Parking their ambulance next to our ARV, one dismounted and the other disappeared into the back to prepare any equipment that might be needed. I joined Cal and the casualty just as he was about to provide a summary to the medic.

'We found her about ten minutes ago, lying in the road.'

The paramedic jumped in. 'Foaming at the mouth, hands all curled up?'

Cal instantly acknowledged the reply. 'That's right. Making some straining noises as she was breathing too.'

The paramedic didn't look particularly interested. 'OK, mate. Let's put her in the back of the ambulance. This isn't the first time I've dealt with her.'

They picked her up and carried her to the ambulance. It occurred to me that a stretcher might be better if she had been hit by a vehicle, but he was the medic.

'Yeah, she goes all around Bristol on her bike and lies on the road waiting for someone to stop and give her attention. There's nothing wrong with her physically. We'll still check her out and call her mum. We've got her number in the control room for when she pulls this kind of shit. Still, we need to check her over just in case she is injured.'

When the mother arrived, she went through her well-rehearsed apologies. Obviously she had met the ambulance crew before, but on this occasion there was an armed police patrol with her daughter as well. Her mother confirmed that she was indeed physically okay and simply refused to walk, preferring instead to use a wheelchair or tricycle to get around. Despite intervention from psychiatrists and other doctors, her mental decline seemed to be progressing, causing no end of stress for her entire family. I felt sorry for them both. And while her act could be easily considered a waste of police time and health service resources, I supposed that her condition was not one of her own choosing. Something was clearly wrong, and her childish behaviour and rapid recovery once in the ambulance also seemed to indicate that perhaps this young woman needed some specialist care that she was not yet receiving.

Her mother picked up the bike and collapsed it in a well-practised movement before storing it in the rear of her car. Her daughter, hand in hand with the paramedic, walked to the passenger side and she got in.

'Once again,' the mother began while closing the doors at the back of the car, 'I am so sorry. I know you have better things to do.'

She looked at her daughter, who was by now sitting happily in the car, adjusting the volume on the CD player.

'That's okay. We didn't mind at all. Drive safe,' I replied and made my way back to the ARV.

Shutting the door, Cal looked at me and started the motor.

'Better things to do?'

The engine came to life.

'Like having a cup of tea I suppose.'

Chuckling to ourselves, we made our way back to Oscar Yankee to begin the paperwork.

# 13

My tour at Oscar Yankee had come to an end. Now I was PC Okuhara, Escort Ops. I liked the sound of that. I had moved to Cheltenham to be closer to work, and I was getting used to living in the town. Having spent most of my life in rural or semi-rural locations, it was nice to have everything I needed a short distance from my front door. Cheltenham had a good feel to it, with wine bars, parks and boutique shops. There was a gym nearby as well, which I decided to join, having spent most of my life training in the great outdoors.

Unfortunately, by now my trainee doctor girlfriend and I had split up owing to the huge demands placed on her during her clinical training. I was seeing a Canadian au pair who worked for a well-paid lawyer based out of the historic Royal Crescent in the town centre. Having heard her accent as she ordered her latte at the local coffee shop, I had struck up a conversation with her and we enjoyed spending the next few hours comparing Vancouver to Cheltenham.

'So what do you do?' she had enquired towards the end of our spontaneous date.

'I'm a police officer,' I replied.

She laughed.

'Is that funny?' I asked, laughing to myself and trying to decide how much of a police officer a CNC police officer really is.

'I've heard that line before,' she noted. 'A lot of guys say that they are police officers or firefighters or soldiers... but usually they sell mobile phones or write insurance.'

I had no idea whether a lot of guys did or did not fraudulently claim to be police officers in the pursuit of romance. However, I did have my badge and warrant card (police identification) with me, so I produced it to prove myself.

'Oh my God! You *are* a police officer!' she said, a little too loudly. I was sure that someone would have heard and assumed that I was making an arrest or carrying out an investigation, or even abusing my position in order to enhance my dating prospects. Talking some more and making plans to meet again, we walked the long way back to Royal Crescent and exchanged numbers.

There was no formal training to become a planner in Escort Ops. The role was so niche and unique to not only the police but also the CNC that it was just a matter of earning experience. Joining the team, I found that I spend most of my time working with another planner and a sergeant. The planner, James Caldwell, had been the shift planner for the southern BCU and an acting sergeant. He had decided not to follow promotion and moved back down to PC rank to become an ops planner. The sergeant, Matt Wheatstone, had previously been the chief constable's driver and after a year with Mr Thompson was allowed his pick of postings anywhere in the constabulary. Not wanting to go back on shift, he had come to Escort Ops on the recommendation of the chief bobby.

James had come to the police after being a grave keeper and had the nickname 'Bones'. He had been on the same recruits' course as Matt, so they were already well acquainted. Matt had been in the RAF as a police officer. Another one. Sometimes it seemed as though there was a constant stream of snowdrops (RAF police) coming and going from the CNC. But that was probably a good thing seeing as their military role was not too dissimilar to our core responsibilities. Matt was half-Australian and had a very dry sense of humour that not everyone could understand, but we hit it off straight away. Even though Bones and Matt were already well established in the police, I was only just beginning my third year. However, we were all starting in Escort Ops with the same knowledge of international police duties: very little.

Ed, the inspector, arrived in the office along with our two civvie administrators. Candice was an American lady in her forties but didn't look a day over twenty. She was fluent in French, so was brought in to assist in planning our operations that involved the French military and police. Our accountant was Amy, who had previously worked for a firm that made satellites for the UK Space Agency. Whereas the police officers in the department had a lot of learning to do, the admins were hugely experienced in their roles. I couldn't help but feel inadequate. We all introduced ourselves and then got down to the first order of business: moving the Escort Ops department from Cumbria to Oxfordshire. It was going to require rental vans and elbow grease.

'Sounds good,' said Matt. 'The three of us are probably qualified to lift boxes and drive trucks.'

'That is a 600 mile round trip,' Bones chipped in. 'So, how do you want to sort it, boss?'

Ed had already decided. 'These two lovely ladies will sort you out with three vans and two nights' accommodation. Convoy up there and convoy back. I want everything moved in one go.'

It seemed tough but not unreasonable.

'We've got some big projects coming up so we need you back and set up as soon as possible. The decorators are coming next week as well. This office is going to be turned into a secure planning centre, so there are no excuses for being late.'

This was a marked departure from my time on patrol. But then again, the role was completely different. I was wearing a shirt and tie instead of body armour and an equipment belt. I was sure it wouldn't take long to get used to this new environment, and the instant issuing of instructions and orders with the prospect of 'big projects' gave me a sense of anticipation and excitement.

'So, planners, start planning.'

With that the meeting ended.

Over the course of the next several days and weeks, the office at headquarters that had been designated for Escort Ops became a secure planning room with separate computer systems, phone lines and coded entry systems. The idea was that during an operation the office would be able to maintain a high level of security concerning the dates of moves, quantities of fuel, personnel involved and intended routes. The entire team had to go through an in-depth security check known as developed vetting. This went so far as to investigate the financial affairs of the officers, their personal relationships and internet search history. Although it was very intrusive, we were always reminded that it was important to be completely honest in all the answers that you gave. I felt certain that my interviewer was very bored with my background. I was still so young that, other than my time in the army, I had done very little. No marriage, no mortgage, no mistress. But in order to be able to access highly classified documents, this kind of clearance is absolutely necessary. The documents and plans within our remit certainly fell within that category.

One of the first tasks assigned to me once we were set up came from the inspector.

'We need you to get onto Dillon Aero in the US and get some naval deck mounts for some M134 miniguns.'

I wrote it down. The previous planning team had ordered the guns but not the deck mounts – meaning that there was nothing to hold them down on the ships. But by now, the previous planners had returned to duty at Sellafield and were not always easy to reach, as they would often be on patrol or otherwise occupied. They were also slightly aggrieved that their ops planning roles had been taken from them so were not inclined to help as much as they could as we tried to navigate our new positions. I already knew a little about the weapon system, however. Dillon Aero manufactures miniguns that fire up to 100 7.62mm rounds per second. When in the airborne role, this rate of fire can be sustained as the cooling effect of headwind and altitude means the barrels do not overheat as quickly. In a naval role, this fire rate can be adjusted to suit the requirements of the user.

The CNC Marine Escort Group (MEG) component provides specially trained AFOs to protect nuclear fuel while it is being transported at sea. In order to protect the vessels (which are operated by Pacific Nuclear Transport and partly owned by the governmental Nuclear Decommissioning Authority), large-calibre and rapid-fire weapons systems are installed on the ships, turning them into armed merchant ships. The MEG operators onboard are trained to use these weapons to defend the vessel as well as recover control of the ship should it be boarded.

I knew that we would have contracts and therefore an account number with Dillon Aero, so I spoke to Amy in order to see what I could find. Within a few seconds she had what I needed.

'Firearms training want some extra batteries as well, so ask about them when you speak to the account manager,' she added.

Miniguns require a power source to spin the barrels and operate the various components of the weapon. Our firearms training team had requested more batteries so that more would be available for the training they had planned.

'Sure, I'll ask,' I said before sitting down at my desk and trying to figure out the time difference between the UK and Arizona.

One phone call later, and after a brief exchange of details with a very helpful American, two deck mounts and two batteries were ordered and on the way to our armoury. I felt quite pleased with myself. It was nice just to be handed work and expected to get on with it. It felt even better to be working in such a unique part of policing. Not so pleased was William Watson. Rather than speaking to me himself, he had instructed Ed to find out 'who the fuck had just spent tens of thousands of pounds on minigun batteries without seeking approval'. It didn't require a detective to find out.

'Matt. A word.'

I made my way to his desk. He didn't offer me a seat.

'Did you order minigun batteries?'

'Yes, sir.'

'Did someone tell you to order minigun batteries?'

'No, sir.'

'Then why did you order minigun batteries when you were only supposed to order deck mounts?'

The first response came out of my mouth without even thinking – the first time this had happened as far as I could recall. Usually I try to choose my words carefully or at least give them some thought.

'Because we need more batteries, sir.'

'I,' he began, with a heavy emphasis on the *I*, 'decide if we need minigun batteries. Not you. Not the training team. Me.'

I could feel myself growing red and uncomfortable.

'And another thing. We can't have two Matts in the department. Come up with a nickname.'

I sat down at my desk and looked around the office. I could see that the other two planners had pretended not to hear what Ed had said to me.

'Coffee,' the other Matt announced rather abruptly. The coffee shop that I had waited in before my interview for the role was only on the other side of the building, and it was getting to eleven o'clock, so the idea of a mid-morning coffee seemed like a good one.

As we queued to be served, a group of CNC recruits led by Sergeant Buckden came into the building through the automatic doors. The chief constable had not only moved Escort Ops to the headquarters in Culham, but also the training school. Apparently his intention was to centralise the entire HQ into one location as opposed to the three it had previously been: Summergrove, Sellafield and Culham. Sergeant Buckden, who enjoyed his position as the training supervisor, had come down during the move and had relocated to the Oxfordshire area to continue his role. They dispersed and sat around several small tables while the sergeant and his new training team disappeared into the secure door separating the headquarters from the coffee shop.

'Good morning sergeant,' one of the recruits announced keenly.

All three of us turned around. I looked at his shoulder number. It was well into the 1000s. Mine was 142. The constabulary had been undergoing a rapid expansion. Starting with Cal and Dave's courses, the training school had been producing recruits non-stop. This was not helped by the high attrition rate of officers. The senior management team were aware of the problem of staff retention but had apparently decided not to do anything about it other than increase the numbers of recruits. Even the deputy chief constable in his opening remarks to this latest course had said, 'Think of this a ten-year job at most.' That could not have been good for morale. I considered myself in for life – at least I had no plans to leave.

'How long you been in training?' I asked.

'It's our first week. We just got our uniforms yesterday.'

The uniforms and equipment had been sent down from stores in Sellafield. That didn't strike me as particularly efficient, but then again, what would I know? The chief bobby owned this train set, not me.

'Well, good. It's a good job and training is great fun.'

Matt nodded in agreement. 'What are you all doing here?' he asked.

Looking at Bones, I could see that he had something in mind. He was barely able to hold his grin.

'We're just having a break and waiting for the training staff to bring us our personal development portfolios.'

Matt nodded. 'Well, we're all drinking coffees. Two black, one with milk and sugar. Here's the money. Bring them over when they're ready.' Handing the correct cash to the recruit, we followed Matt to a mezzanine that overlooked the coffee shop. The recruit followed us with his eyes, confirming our location then went about ordering the refreshments.

'Right then,' he said as we all sat down. 'Minigun batteries... You prat. That's around £20,000 you spent and we can't get a refund.'

I considered going through the same reasoning that I had with Ed but just stayed quiet, sending a little nod to our recruit/waiter to indicate we were ready for our drinks.

'But, we were going to order them anyway I suppose. Just not from this budget. We don't have the procurement processes in place after the move and it's stressing the boss out. It's not your fault.'

I understood what he was getting at.

'Any high value purchases should come through me next time, and I will pass it up the chain. That's how things work. That's why we have accountants.'

It was reassuring to hear. Some support and advice from the sarge and a free coffee to boot.

# 14

Maritime operation callsigns for the escort group were named after cars. Escorts to Japan were named after Japanese cars, and so on for Europe and the United States. Operation Accord was an escort of spent fuel on behalf of the French nuclear industry to Kobe, Japan.

It was 2008 and this was the first international policing operation for the CNC in almost a decade. In the interim, the MEG and REG had only existed as a trained body of operators that had not carried out any escort activities. While not on operations, the officers who made up the group were seconded to other units, carrying out normal CNC duties and periodically requalified in their specialist skills. Most of them were based at Sellafield and Dounreay, the largest units, but others, like Aled (who was a member of the REG), were at the smaller power station units and sites. Some of the older members had taken part in fuel moves to and from Japan in 2001, but there were also a number of officers who had not even been to sea outside of a training environment. There were also new officers to the group who needed to undergo a full training package before they could be deployed. As ops planners, Bones, Matt and I had to come up with a scheme of work to train the new officers while at the same time running pre-embarkation training for the rest of the MEG.

Carrying out a firearms operation of this scale is not as simple as just putting some well-armed officers aboard a ship and waving them goodbye. The workup training, qualifications and administrative actions can take months. For example, each ship will require a certain amount of ammunition. This is not a typical 'police' load of ammunition but rather enough to supply the various miniguns and general-purpose machine guns (GPMG) and other weapon systems aboard the ships. There has to be enough for continuous training, exercises and hostile situations – not an insignificant number of bullets. These bullets need to be ordered from British Aerospace, produced, and then sent to the CNC ammunition bunker at a confidential location. Then the crates of 7.62mm need to be collected by the planners and taken to either the ships or the gunnery range. This all takes time.

Many such actions need to be carried out. The ops planning sergeant therefore had the PAR (plan/action register), a checklist of thousands of items that needed to be completed before an operation could be carried out. Matt would assign tasks to me, Bones, Candice and Amy depending on the subject.

Matt and I were on our way to the Ministry of Defence location to collect ammunition for the GPMG qualification shoots that were taking place at a gunnery range in Wales. Guarded by the MOD police, this massive site had its own train network to move the various types of ordnance and military equipment that was being stored there.

'You used to be in the infantry,' said Matt. 'How long do you think it'll take to move all those boxes of ammo?'

Between the two of us we had to collect enough bullets for two shiploads of officers and the reserve crews.

'Well... it depends on how close we can park to the bunker. I guess an hour or so. Didn't you use gimpies[1] in the air force?'

1 GPMGs are often referred to as 'gimpies'.

'Nah, we had them but never used them. Noisy, dirty things that take forever to clean,' he replied.

He had a good point. The machine gun worked using gas recoil, meaning that as the bullet was fired some of the gas (and byproducts of shooting) was recycled back down the weapon to keep the gun firing. This created a thick coating of black carbon on the working parts of the weapon.

Arriving with the ammunition at the range, I was surprised to see that Chris, my IFC firearms instructor, had made a sideways move into Escort Ops and was now an instructor for the entire group, covering marine and road escort tactics.

'I heard you were coming,' he said as he extended his hand, offering me a warm welcome. I had no idea which of the blue gods were allocated to us, so I was surprised and pleased to see Chris, who had made such a great effort to help me through my first few months in the CNC. Without a doubt, working with experienced and motivated Escort Ops officers was a good move for him, and he genuinely seemed happy to be running the GPMG course. He had even convinced Jacko to transfer across as well. We got to it, and Jacko helped Matt and I unload the bullets and store them in the ammunition bunker. The officers on the course would be the ones unpacking and preparing it for firing – which gave us time for a brew and a catch-up, as well as the chance to talk about the operation. 'I think this job'll suit you,' remarked Chris. 'More to think about; problems to solve. Not just standing around at gates watching people come and go.'

I agreed with him. I cast my mind back to Oldbury, where I had done exactly that. I did miss Oscar Yankee, though, and my colleagues there. However, within just a few weeks of joining Escort Ops I had travelled around to various nuclear sites, become involved with high-end training and been chewed out for ordering unauthorised minigun batteries. It was already proving to be more interesting than two years on the beat had been.

We went into a cabin adjacent to the range loading area.

'This one's called Op Accord,' Matt said as he produced a laptop with a digital version of the PAR on it. 'We don't know the exact date for deployment yet, but the assumption from Wally and the governor is that we are on a tight timeframe.' Jacko and Chris moved behind Matt's shoulders to get a glimpse at the never-ending list of tasks on the screen. 'So we're going to need to allocate you some firearms training, specific tasks. If it's something you cannot do let me know now so that we can get as much of this squared away as possible.'

Running through the list, Jacko and Chris gave a curt yes or no to each task that was thrown their way. Both instructors had been on the 2001 move and knew more about planning than the planners. They knew that we were still trying to find our feet and were willing to help where they could – to the extent that they were brutally honest with the sergeant about how much the two of them could do without bringing in extra manpower.

'Alright, Matt... sarge, not that Matt,' Chris said, jerking his thumb in my direction. 'We know what needs to be done and most of what you have on that list just happens organically anyway. However, just how much do you expect the two of us to do while running pre-embarkation at the same time? Have you thought about bringing in some extra bods to make this more workable?'

Matt didn't look up from the screen, analysing Chris's comment and task allocation at the same time.

'Yeah.' A response that didn't really answer the question. I hadn't known him long, but it seemed that Matt had two distinct personalities that never overlapped: fun Matt and work Matt. 'Yeah' again. 'I'll pull a couple of guys from Sellafield firearms and that will give you four instead of two. Let me know who you want and I'll get it sorted.'

The laptop computer closed and the sarge stood up.

'As for now, we need to head back to HQ and see what else is waiting for us. Drop an email to me and copy in the boss, when you've decided who you want.'

Suddenly I felt like a spectator. These two instructors had been in the police for years and knew the score better than either of us. Matt, the chief constable's former driver, had switched in an instant from friendly boss to serious, thinking ahead to the conclusion of an operation he had only just been assigned. We knew that he had no authority to redeploy people, but I could also see that he knew he could get it done.

Back at HQ, we found that the planning phase of the operation was moving on nicely. In our absence, Bones and the rest of the planning team had quickly taken care of a massive section of the PAR, mostly involving administration and staffing issues. The two teams of officers required for the escort had been selected: a team for each ship. There were still two teams of reserves to put together and a chief inspector to select as the operational commander on board the ships. One of the teams was going to be sailing on the *Pacific Heron* – a ship that had only just been commissioned and was not yet fully operational. The others would be on the older sister ship, the *Pacific Pintail*, which was going to complete its last voyage this time around. Outwardly the two ships looked identical, but the *Pintail* was much more dated. The fleet of ships owned by PNTL (Pacific Nuclear Transport Ltd)[2] were all custom built at huge expense as nuclear material carriers and as such had specific capabilities and systems that made them suited to the role of long voyages carrying armed police and potentially very volatile cargo.

2 PNTL is a shipping company, majority owned by the UK government, that specialises in the secure transport of nuclear material.

Ed came into the planning room following a meeting with the senior management team including the chief constable.

'We've got a date to work towards,' he announced, closing the door so that nobody could eavesdrop on the briefing. Ed always had the decency to stand up as he talked to his team, and never gave a briefing from behind his desk.

'Now that you know the date, I think there may be some things on the PAR that we cannot achieve. I'm mostly concerned about kit and equipment. So Matt and... Matt. Matt, do you have a nickname yet?'

'No boss.'

'Well hurry up,' he ordered. 'Next week you need to go to Barrow and meet up with the *Heron*. The crew will be on board to start their training. Familiarisation and things like that. Get a list of everything that they need and then hire a van – pick it up from Sellafield and drop it to them. Same again for *Pintail*. It'll take you all week. Speak to Candice to get a decent motor and some accommodation. Matt... the sergeant Matt; you need to get up there as well to speak to the Inspectors so that we can get some specifics for the operational order to be written. Bones, you hold the fort here and work with our two ladies to make sure that the admin is silk as milk. I'm going to HMS *Collingwood* to meet up with the Navy and then to Poole to discuss some details with the other people. Any questions?'

We all looked around, having been bombarded with a mass of information. But we had nothing to add. We were here to solve problems and answer questions, not ask them.

'I've booked you both into the Abbey House,' Candice said in her soft American accent.

'Sounds good,' I replied, taking the envelope with the travel details.

'I didn't book a car,' she said. 'You can take one of the unmarked police cars; they need a decent run.' I liked the sound of that. Escort Ops retained some powerful V8 Land Rovers for road escort reconnaissance work, but often they were static in the constabulary car park if not being used for training.

'You can stay on the ship if you prefer,' she added as I walked back to my desk.

'No thanks… I think the Abbey House sounds more like my thing. What about a truck or van? Can you sort one of those out?'

She pointed at the envelope. 'It's all in there. Have a nice trip.'

Once again I found myself driving from the south to the north – but this time I had brought a 'learn French' CD to help keep me occupied between service stations. The V8 engine was very thirsty; I could almost see the fuel gauge move. Checking into the hotel, I saw that Matt had already arrived in the other Land Rover. The receptionist explained to me that there was a wedding going on, so part of the hotel was reserved for the party and guests. It also meant that the hotel was at maximum occupancy, so I had been put into a suite instead of a regular room. I looked at the key fob that she handed me. *The Princess Suite* was written in flowing calligraphy across a stained oak key fob.

Heading into the suite, I was impressed with my luck. There was a three-piece leather suite around a fireplace and a flatscreen television, a dining table with four set places and a four-poster bed complete with lace curtains. I checked the bathroom – a jacuzzi. Jackpot. Certainly good enough for a princess, I supposed. If it was going to be a busy week, at least I would be able to return to a certain amount of opulence in my time off.

Down at the bar I met up with Matt, who was writing some notes while enjoying a pint. It was starting to get late.

'Alright sarge? What's that?' I asked, looking at his full pages of notes. He must have been working for hours.

'The op order notes I have so far. I'll write some ideas down here then type it up with the inspectors from *Heron* and *Pintail* later.'

I nodded in understanding. 'You want anything from the bar?'

'Yeah, I'll have another,' he said. He wasn't halfway through his first drink but I guessed the Aussie in him wasn't about to turn down a free drink.

Returning to the table, drinks in hand, I set them down and picked up some of his notes. 'Bloody hell, mate. There's some stuff on here.'

'Tell me about it,' he replied. 'The super needs to see the draft by the end of the week, so we're both gonna be busy. Are you sorted for visiting the *Heron*?'

I nodded while taking a gulp. 'Yeah. Never been on board before, and never met the boys before so it'll be something new.'

He chuckled in agreement. 'Don't let them take the piss with equipment requests,' he said while putting down his pen.

I knew what he meant. The escort officers were entitled to equipment far and above the general issue, but just like the military they had a tendency to seek out shiny new things, particularly for 'trial purposes' or 'because they had lost their old one'.

'I'm going to finish this in my room,' Matt announced, scooping up his notes and fresh pint.

'I'll give you a hand. You dictate and I'll type it out.' I had learnt to touch type at school, so was quite fast when it came to writing on a computer. 'We'll be done in no time. Plus I've got a table in my room we can use to set up the laptop and printer. I'm staying in one of the suites so there's plenty of space.'

'Shit, this is a nice room.' Matt went back out and checked the name of the suite, which was on the door. I hadn't heard him swear before. 'The Princess Suite, eh? Well, like it or not from now on you're called Princess. I'll email Ed now and let him know.' Half joking and half serious, he produced his Blackberry mobile phone and began typing away.

'Really?' I asked.

'Oh yeah. The best nicknames have a story and the high-strung Princess needs his suite, four-poster bed with net curtains... I bet you ordered room service too.'

I had been planning on it but decided not to comment. It wasn't the most offensive nickname I had heard given out in the CNC, so Princess it was.

The draft notes were finished sooner than we had thought, but they were also shorter than Matt had intended them to be.

'This is going to need some work,' he said, taking the freshly printed pages from the portable printer. 'I'll go over it again in my room, now. I'll use the captain's dayroom[3] on the *Pintail*, which is moored alongside the *Heron*, to rewrite it. If you need anything just come across. If I don't see you then meet me like today, and we need to get this squared away. See you later.'

Matt left the suite, notebook in hand and with the encryption key needed to activate the laptop. I guessed that any final amendments to the op order would be carried out here, where he had left the printer and computer.

Being off duty, I decided to call my girlfriend. I hadn't seen as much of her as I was hoping; Operation Accord was taking up a lot of my time, but we had been able to speak regularly at least.

'Hi,' she purred down the phone. 'Did you have a safe trip?'

I told her about the hotel and suite.

'I could never afford a room like that. A four-poster bed and fireplace? A princess's suite?' she giggled. 'Can I come and visit?'

---

3 Unlike the captain's cabin where the captain spends their off-duty hours, the captain's dayroom is where the captain can be found when not on the bridge.

I didn't know if she was serious or not. Cheltenham to Barrow-in-Furness was a long way, and she still had her au pair job to do.

'I don't mind driving,' she added seductively, 'and I've only ever seen Cheltenham. While you're at work I can see the Lake District and then we can spend the night together and… things. I'll make it worth your while.'

Hard to argue with that.

# 15

Simultaneously with the planning of Operation Accord, the officers responsible for the armed escort were conducting their pre-embarkation training. The comprehensive package of maritime security skills was overseen not only by the police but also by the Royal Navy, specifically FOST (Flag Officer Sea Training),[1] the branch of the military overseeing competence to operate at sea. We were aware that a final exercise would need to take place before the ships deployed with their nuclear cargoes, so it was important to address as much of the PAR as possible before that time. There was also a requirement to engage with some of the other agencies regarding their involvement in the operation. Cumbria Constabulary would provide overt unarmed policing and divers. Additionally, there would be specialists from the military on standby, though they would not be visible during the UK phase of the operation.

Ed had enlisted me as his driver to take him to HMNB *Portsmouth* to attend a meeting between him, two FOST officers, and other military personnel directly involved with the escort.

---

1 FOST was renamed to Fleet Operational Standards and Training in 2020.

'I'm relieved that we've sorted out that nickname,' he remarked from the backseat as I drove the Land Rover south down the M3 motorway. 'Having two Matts in the office isn't ideal, and it's quite confusing. He outranks you as well, so that's settled, Princess.'

He was perusing files and checking notes on his Blackberry. I could sense that he wasn't in the mood for a conversation, just making casual talk. Despite his demanding role, Ed could at least engage in small talk with the officers in his team. However, his attention was focused on the upcoming meeting.

'Should I join the meeting, boss, or wait with the vehicle?' I inquired.

'Oh, you'll be in the meeting, no doubt. I won't be taking any notes. You're my driver and secretary today.'

I was fine with that arrangement. Even though taking notes could be a tedious and frustrating task, it would allow me to learn more about the military's involvement, of which I had little knowledge until now.

Ed received the notification from superintendent Watson as we drove through the gates to the naval base.

'Princess?'

'Yes, sir?'

'Are you current?'

'Yes, sir.'

'After this meeting, you'll be training for marine and road escorts. We're short-handed and have a new task. You need to organise your own training and need to slot in with the buildup training where you can. Run it by Bones – I'm appointing him acting sergeant and ops planner for a road escort task in Dounreay that just came up. If anyone gives you shit for not going through the normal selection pathway to get on the team, direct them to me.'

That caught me off guard. But I was also aware that, as a police officer, time off could be subject to the operational needs of the constabulary.

'Do you want me to work on both operations, boss?' I inquired, expecting a heavy workload.

'Yes. The super just made Matt up to acting inspector to run the road escort as well.'

I wasn't disappointed by the news. Both of my fellow planners had just been instantly promoted and I had been given the opportunity to take part in some serious policing. When I joined Escort Ops, I had expected a busy role involving creating plans and arranging logistics, but now, with the possibility of armed deployments, my role had evolved from administrative work to being actively involved in escorts.

Back at HQ in Culham, the rest of the team had received a notification about a new operation. The Americans needed to transport a specific isotope of nuclear material from Dounreay to the United States, utilising a military transport aircraft based out of Ramstein in Germany. This kind of operation needed less planning compared to a sea transport, but with the department's resources already stretched to the limit, Bones decided to concentrate entirely on preparing the groundwork so that as soon as the ships were at sea we could be prepared to carry out this short-notice operation.

The meeting with the Navy revolved around the assessment process for the two crews and reserve teams while at sea. They needed an understanding of the operation so that they could plan some realistic scenarios. Also in the meeting was a major from the Royal Marines. The style of the parachute wings on his uniform indicated that he was probably more than the humble operations officer that he claimed to be.

'I understand you will be using a new ship, Inspector,' he added while the FOST officers were collating their notes. 'We need you to take us on a tour of the whole vessel.' He never mentioned who 'us' was.

The *Heron* had recently been introduced to the PNTL fleet and was so new that some of the interior had not been painted.

'We will also need a few dates for some interoperability training on the *Sir Tristram*[2] as well. Within reason, you can pick any day in the near future.'

I was busy making notes as the requests were coming in thick and fast. No doubt a lot of the questions that could not be answered immediately would need to be answered soon. But it was hard to keep my mind from wandering to the prospect of deployment. Even as an operations planner, I had maintained my AFO status – but now I was busy thinking about what training I could place myself on. Selfishly, I was also thinking about what training would be the most interesting as opposed to the most relevant. It would be possible, for example, to go to sea without a 30mm cannon qualification or an emergency dentistry course, but a sea survival certificate was essential.

At the time, most recruits to the escort group were placed into either the MEG or REG element, although there were some officers who were qualified in both areas. In addition, the firearms training team were qualified to instruct on escort-specific weapons and training scenarios. For this they often needed to join the navy or other military outfits to become qualified. The joining process for most of the officers began with a paper application, followed by interviews, shooting tests and fitness tests. If these were all completed satisfactorily, the highest-scoring applicants were offered a position on the team. This meant that, even without escorts to run, the CNC always had a cadre of escort officers available. With a lull in international fuel movements, however, many of these officers were only partly trained and partly equipped. Luckily for me, this meant it was relatively straightforward to join in the training as most of the courses would have at least a few novices.

2 *Sir Tristram* is a large, decommissioned transport ship that is regularly used for training special forces and marines.

I was also thinking about equipment. I had a pretty solid understanding of what I would need, as I was already handling the clothing and equipment requirements on the PAR. Over and above the standard policing equipment, marine or road escort officers carried specialist equipment that was particular to their role. Marine escorts would have a lot of flame-retardant equipment in case of emergency aboard the ships, as well as sea survival gear. Road escorts had helmet-mounted night vision and additional communications gear for use in open country areas. My first stop before training would be the stores back at the mothership, Sellafield, in order to get what I needed.

# 16

I went to stores at B357 and met Marie. We had been in touch regularly by phone and email – not socially, but because I was handling all the clothing and equipment requirements for the department. The shelves were piled high with gear for the entire constabulary, including items of UKAEA vintage that might become useful again one day. The shelves contained everything that an officer could need, and as we were busy putting together a kitbag of equipment for my upcoming training I couldn't help but explore a little. There were cycle helmets from a failed concept that would have seen unarmed officers patrolling Sellafield on mountain bikes; there was head-to-toe body armour for riot control. There were stethoscopes and comprehensive first-aid kits, and night-vision equipment, respirators… anything that a copper could ever want. No wonder only the stores personnel could come in here. Even the most honest of police officers might suddenly decide that they could benefit from some additional hardware.

'So you'll need to try on some coveralls – you want them quite loose fitting so they don't stick and rub when you get hot. You'll need a bush hat and some lightweight boots…'

Marie was thinking out loud as opposed to talking to me, all the while filling up the kitbag that I was carrying. I was aware that it was getting pretty heavy but I didn't say anything. I didn't want to make a bad impression with Marie, who seemed to know the entire constabulary.

'Have you ever been to sea?' she asked, placing a riot helmet and visor into my overflowing kit bag.

'No, never. Well, I went on a cross-Channel ferry a few times.'

She laughed. 'Well, that's going to be an experience for you,' she commented. 'I hear you either love it or hate it. The ones who love it don't get seasick. The ones that hate it do get seasick. I hope you don't get seasick.'

I hoped that too.

'One officer was seasick every day of the voyage the last time they went out to Japan. They were going to drop him off before collecting the fuel but he insisted that he would stay. There's no way off once the ship is out in the ocean... as far as I know that only happened once. One lad broke his collarbone in a storm and an Australian Navy vessel came to send a helicopter to take him off – but that was three weeks after he'd broken it!'

Encyclopaedia of knowledge that she was, I decided to take advantage of Marie's willingness to share marine escort stories.

'On a run out to Charleston in the US, Sebastian, who is a vegetarian like you, had macaroni cheese every day because the stewards couldn't be bothered making him a separate meal,' she recalled. 'So long as you don't get sea sick and so long as you like mac and cheese, everything will be great.'

She took the bag from me, demonstrating not inconsiderable strength, and dropped it on a worktop.

'Okay. Now for some road escort gear.'

I arrived at West Yorkshire Police's accommodation block in Wakefield just as the rest of the MEG were turning up on a

Monday evening. Our own constabulary firearms training unit was oversubscribed with recruits and firearms units, so the decision was made to train the escort group at a different location.

Parking my car and carrying all my equipment inside, I could feel the eyes of the team on me. They had already started their training a couple of weeks ago – working on fitness and spending time on the ships to become familiar with their layout and non-police duties once aboard. I felt like an interloper, and I could hear a few murmurs. They knew who I was, though, so I wasn't a total stranger. While they all helped each other carry in bags, boxes and crates, I was left alone to sort myself out. I thought it might be a good idea to help them with their equipment once mine was inside and introduce myself properly, but by the time I had found my room and put my gear away they had finished as well.

I joined the Operation Accord officers in the lounge that connected the east and west parts of the seventh floor of the accommodation block. It felt like an outdated block of flats. It could well have previously been part of a university halls of residence or a budget hotel at some point. The team was spread around a large and tired-looking common room, using the kitchen area to make tea. Some were watching TV, others were chatting in groups and a few had boarded the lift, intent on having a smoke break outside the ground-floor entrance. We were waiting to be joined by the training team, who were travelling from Sellafield with two additional instructors. They were going to lay out the week ahead and give us a heads up on the schedule – which seemed to be quite fast paced. I already had a notebook full of the details and what we were doing. Some of it I had helped arrange.

'Oi, Princess,' someone called out. 'Princess, isn't it?'

I turned to the group in the kitchen to see a sergeant and a few other officers, cups in hand and eyes directed at me.

'Matt said you were coming.' The sarge put his cup onto the kitchen side and extended his hand.

'I'm Danny, this is Kev and he's Jimmy Guns. I'll introduce the others after they've had a smoke. You'll be working in my team.'

Danny was one of the reserve team's sergeants. He was an older officer in his forties with grey hair and very dark eyes and what seemed like a permanent grin on his face, as if he was trying to stop himself from laughing. He was the shortest of the group but still looked very powerful, like a swimmer or triathlete might. He had been on the previous deployment to Japan as a PC and was now training his team to act as replacements where needed and as gangway security while the ships were in port. Kev, one of the PCs in his team, was younger than the other two in the group. He spoke with a Cumbrian accent and had been with the MEG for only a year. He worked at Sellafield and was very popular. He was often ready with some anecdotes from his personal life. Whether they were accurate was another matter. Finally, the other officer, Jimmy Guns, was also new to the team like me and obviously liked to train with weights. His arms were conspicuously large, earning him the nickname 'Guns'.

I tried to introduce myself to the team but they seemed to already know enough about me to make it unnecessary for now. Kev had been on the same initial training course at Summergrove with Bones and Matt, so of course he briefed everyone accordingly about the new arrival to the group. I had bumped into a few of them before while carrying out equipment requests and admin onboard the ships as well but had never been properly introduced. As the ship's crew were guiding the security detail around the ships, describing the systems, procedures and actions they were expected to perform, I had been carrying my notebook and running around taking equipment orders to bring back to Marie at stores.

Not long after I had tried to settle in, the firearms trainers emerged from the lift and joined us in the common room. They were also staying with us, in the same building but on a different floor – no doubt so they could discuss student performance in private. Chris and Jacko also brought with them a training sergeant called Glenn and another PC called Wynn. Matt had managed to get some additional people released to conduct the training as promised. We gathered around to hear about what to expect over the next few days.

'First off, introductions,' said Chris. 'Glenn is on loan to us from firearms training for this part of the course only. He's looking to gain some experience for his promotion board so be nice and make sure he goes home in one piece.' Glenn was a tall former marine who didn't seem to say very much. Maybe he was waiting for the right time to speak.

'I'm sure you all know Winnie' – everyone jeered and laughed. A long-time instructor, most officers had come into contact with Wynn at some point and were by now used to his rough but efficient way of training.

'Finally, there's young Matt from the planning team.' Everyone turned to look at me as I stood with the reserve crew. There were a few nods and gestures in my direction. 'As you know, Nick has had a bereavement that requires his immediate attention and had to leave the reserve team… we're still early in this phase of the job so Princess has agreed to step in and take his place.'

Jacko took over and outlined the scheme of training that they intended to follow.

'Alright boys, it's nothing new. Those of you who only need to requalify will still be going through the entire package along with the new guys. We're starting with G36 tomorrow and see how we go from there. The Glock, well, that's a piece of piss so those of you who haven't used it in the escorts configuration before – don't worry.'

The team were all paying attention the four instructors. Even though the highest rank among them was a sergeant at this stage of the operation, they were responsible for all the escort team training, ensuring we were suitably qualified to be deployed.

This was more about function than rank, as the team included inspectors and we would soon be assigned a chief inspector as well. These senior officers needed to undergo the same training as the rest of the team – rank meant very little at this stage.

I had previously fired automatic weaponry with the military so was not too concerned about the upcoming training. The G36s that we carried could fire single-shot, three-round bursts or fully automatic, making it easier to engage boats that might be running alongside the ships. In October 2000, the USS *Cole* had been attacked by fast-moving boats that had suicide bombers onboard. Seventeen sailors were killed and the ship was crippled, showing the damage explosives can do if they get close to even an armoured hull. This was less than a year before the attack on the World Trade Centre, indicating an alarming rise in suicide attacks and well-coordinated terrorist organisations. It was determined by naval forces all over the world that well-armed gangway security and deck patrols would be able to offer an effective countermeasure to this kind of attack. Providing 360 degrees of protection at all times meant that, unlike most AFOs in the UK, CNC officers on marine escort duties have automatic weapons available to them.

We also had to review our pistol work with the Glock 17, the difference this time being the weight of a powerful torch mounted underneath the barrel and a larger holster to accommodate the firearm. These guns were needed for use inside the ships, where the long barrels of the G36 were less practical. On the previous deployment to Japan the escort group had brought shotguns, but it had been decided by someone in an executive position that shotguns were not ideal and potentially too dangerous.

'They're guns,' said Kenny as he told me his thoughts about the decision. 'They're supposed to go bang and do the business.'

Kenny had been in the police for over twenty years at this point and, like Danny, had been on the first operation to Japan. He was a Scottish officer based at Dounreay. For training he usually wore his initial issue of firearms equipment that he received when he had joined the UKAEAC. His coveralls had a large brass zip and his helmet looked like something from the Second World War. Even his leather gun belt looked tired, but he reasoned it was more comfortable and so kept his operational kit in his bunk space on the ship. Other officers, too, at this stage of the training were wearing their regular patrol gear on the range. It was what they were used to and there was no hard and fast rule about wearing escort group equipment while in pre-embarkation.

The week progressed smoothly. The range at West Yorkshire was modern, and certainly better than the one at Sellafield. It had more lanes, meaning more officers could shoot at any one time. I found myself enjoying the training. It felt different to being a recruit. Two years ago, I was shooting to earn a place in the CNC. Now I was shooting to join an international policing operation. We fired our weapons at multiple targets, from different distances and from cover. We fired as teams and individuals. We were getting through hundreds of rounds each day. Within a few days, everyone was signed off as competent on the individual weapon systems, meaning that the rest of the range package could be spent training for tactical scenarios.

The following week we headed out to practise and qualify in using the support weapons – the miniguns and GPMGs. These guns could not be fired indoors, so we had to use a military training area instead. I had been to Otterburn before, more than once. This large training area had been used since the First World War as a venue to practise battle manoeuvres. Most of the team only needed to requalify, but some of us needed to start from the beginning. I was already familiar with the weapon system as it had been in use with the British military since the 1960s, but I had never used it in a naval or policing context before.

'This is the L7A2 GPMG,' said Chris, as we stood around a table in a classroom in a nearby army camp. 'I know that you may have shot this before – but this is a unique weapon as far as the police is concerned so that means there is a strict procedure for training in its use.' Here he looked at me and Jimmy Guns. 'Which means even if you think you know better than me; you don't. The GPMG is a belt-fed, air-cooled, gas-operated machine gun that fires the 7.62mm NATO cartridge. It's known for its reliability, durability, and versatility. It does the business.'

Our course was only four officers – the rest of the team were with Jacko, Wynn and Glenn, taking it in turns disintegrating the Northumbrian countryside, 200 rounds at a time. Along with Jimmy Guns and me were Craig and Dillon. They were also assigned as reserves and were newcomers to Escort Ops. Like a lot of the team, they were based at Sellafield. Craig was a keen amateur footballer and played for Whitehaven FC when he wasn't on duty. Apparently he was very good. He certainly looked like a footballer with his trendy haircut. I guessed he had modelled it on some famous player. Unfortunately, the rest of him did not look so trendy as he had a prominent overbite and acne scars on his cheeks. Dillon was the oldest of our group, and in his mid-thirties. His short, wide frame made him almost box-shaped. But he had plenty of charisma and was popular with the entire team.

Jimmy had previously served in the infantry and was already complaining about the impending and guaranteed cleaning of the gimpie. Chris went onto explain, pointing at the various parts of the machine gun with a pen as he spoke.

'This is a gas-operated machine gun, meaning that some of the gas generated from firing the bullet is recycled to allow the piston to operate the feed mechanism – which will in turn continue to operate until you release the trigger or run out of bullets.'

It was a simple explanation and one that I had heard before. Craig and Dillon, however, were getting their hands on this serious ordnance for the first time, and keenly examining the GPMG.

Picking it up, Craig asked, 'How heavy is it?'

It wasn't overly heavy at around 15 kilograms, including a short belt of bullets, but it was noticeably heavier than anything else these officers would have used.

'It's mounted at fixed points on the ship but can be quickly removed and positioned anywhere. That being said, you will need to learn how to operate this thing both in the mounted and light role configurations,' concluded Chris.

We had one gun between the four of us, along with a supply of ammunition that Matt and I had moved from the bunker a few weeks earlier. I could hear the other guns on the training area, firing short bursts. I recognised the deep, rapid bursts and following thuds as the rounds landed on the target. It was going to be a long day for the other officers, who would finish the morning shoot, then return at night to practise firing with NVGs (night-vision goggles). We, on the other hand, were going to be guided through operating this belt-fed automatic weapon as if we were in the initial stages of basic training all over again.

Whereas the course was split between initial training and refresher training for the GPMG, all of us were using the minigun for the first time, with the exception of the instructors, who had been trained by the Royal Navy off Gibraltar, where they also undertook their 30mm cannon refresher at the same time. For training, we only had two guns. The others were being fitted to the ships. Instead of having a full ship's company gathered around the guns, the teams were further divided. The PCs who would be carrying out most of the duties were split into one group. The sergeants, inspectors and reserve officers made up another group. Those who were not carrying out training would be assisting in the running of the course. We may have only had two guns with us, but we needed a lot of ammunition: 6,000 rounds per officer. This all needed to be loaded correctly and prepared for firing. As the magazine for the minigun can hold thousands of rounds, we needed to link several belts of ammunition together and set it into the magazine correctly. Loading just one magazine took a very long time.

After a full day of classroom work, theory, capability study and taking apart, lubricating and reassembling the guns, we made our way to one of the many hills that Otterburn uses for gunnery training. Using all-terrain vehicles, we took everything that we needed to the firing point, then returned to bring up any officer who could not fit on during the first trip. It reminded me of the old Second World War footage where bomber crews would ride on overcrowded jeeps and bomb trolleys to get to their aircraft. We looked similar. Most of the officers were wearing fleece-lined winter uniforms, designed to protect the wearer from the worst weather that the ships might experience during the operation.

We set the guns onto the mounts at the top of the hill. Far below us, a kilometre away, was a target frame that had been brought by the constabulary armourer, Gibbo. He was here just in case there was a serious issue with the guns. Knowing that I was from the planning team, he kept insisting that I ask the powers that be to send him to Dillon Aero in America so that he could become more familiar with the minigun.

'I can't just do this based on some basic experience and old memories,' he reasoned. 'You need to send me out there to do a full course.'

Well aware that he was more interested in a trip to Arizona, Wally had declined his request several times and was considering sending him to a naval base in Scotland to carry out the training if he kept insisting.

'Sure, Gibbo, I'll let them know,' I reassured him. It wasn't much to do with me anyway. My mind was on getting to grips with the training.

The first magazines had been loaded the night before, so those who were not shooting had nothing to do while the instructors took the first officers to the guns. We knew that the guns were going to be loud and that they would soon be glowing red hot with the heat of the bullets being spat out of the barrels, which rotated at high speed, powered by the batteries I had bought without permission some months earlier. On pressing the button at the back of the spade grips, the gun roared into life as the first two officers watched their tracer bullets fall. The guns were also mounted with holographic optics, allowing the gunner to have better accuracy while firing. However, at such a high volume of fire, accuracy was replaced in large part by a lethal wall of lead that would shred any targets in its path. It was extremely loud. The tracer rounds almost created a solid line from the gun to the target. It looked a bit like a laser.

Burst after burst was fired as each officer became familiar with the weapon's handling and weight. Looking under the large ejection chute I could see hundreds of empty bullet cases landing around Jacko's feet. He was standing off to one side, checking on the gunner and not the accuracy of the shooting serial. Within seconds the barrels were becoming red. After less than a minute they were glowing. Once the magazine was empty, the gunner produced a large screwdriver to rotate the barrel, allowing the working parts to spit out any cases or link that were still in the mechanism. If he touched it by hand, no doubt he would be giving up his place on the crew and a reserve would need to step in. It was impressive to see. I had never seen firepower of this kind that wasn't attached to a tank or helicopter.

When my time came, I took a hold of the weapon grips and adjusted my position so that I could see through the holographic gunsight. The body position was completely different to anything I was used to. Thinking about my old instructors in the army, I could almost hear them saying, 'Butt in the shoulder, good eye relief and a gentle squeeze of the trigger.' Not with the minigun. I stood back and placed the red dot over the target. Instead of pulling a trigger I pressed a button and the weapon burst into life.

The torque of the barrels required me to return the stream of fire to my target. I only had 3,000 rounds. Only 3,000? That was one minute of ammunition. Jacko was shouting something at me as I brought the tracers to bear, no doubt causing devastation to the target area.

'Yeeeeeesss,' he called. 'Again!'

I opened fire. The gun burst into life. It was such a powerful weapon. I could understand why it was so useful to the escort group. Any fast-moving attack craft that threatened our ships would be blown apart in no time.

# 17

Between training courses, most officers were able to take leave and spend some time away from the job. I on the other hand had to return to HQ to carry on arranging the operation with Matt and Bones. Back at Escort Ops, the chief inspector had been decided for the operation. It was Ed. He had been temporarily promoted only days earlier, owing to a lack of chief inspectors willing to go to sea. As strategic commander of Operation Accord, he was expected to coordinate every aspect from aboard one of the ships. As this would require him to be at sea along with the rest of the officers, he was also going to need to take part in certain parts of the training, most significantly sea survival and mandatory maritime skills.

I had just arranged at short notice a course for him, two other officers and myself to attend at a private maritime college. Usually the sea survival course was outsourced to Northumbria Constabulary, where officers would train in the River Tyne and the North Sea. Over the course of a week, escort group officers would learn skills essential to emergency situations, such as abandoning the ship, using lifeboats and operating the various aids that could be used should the worst happen. Ed wasn't keen on the idea of getting cold and wet, so he asked me to find a training course that used a heated pool instead.

The college I found usually provided training for crews who were expecting to work in the Mediterranean, providing services to super-rich yacht owners. Arriving at the college we were greeted by some of the staff, who opened the door to our unmarked police Land Rover and invited Ed into the building along with the other two officers. Being a chauffeur for the time being, I watched them go inside and parked the car before joining them at reception. The college director had come down to meet us, no doubt happy that he was now able to add that his training had diversified into government agencies. He was asking all sorts of questions about the nature of our work – which of course we could not answer. How many are on the crew? How long will you be at sea? Where will you be sailing?

Ed guessed that he was fishing for business and not a spy, so he provided him with some vague answers and changed the subject to what we were expecting to do. The rest of the escort group had already undertaken the week-long course, which was specially designed for police officers who work at sea, with the Ministry of Defence Police, Metropolitan Police and Hampshire Police Marine Support Units also being regular customers. With limited time, however, we were doing our course over two days and indoors, in a heated pool, as requested by the boss.

'Good work Princess,' Ed noted as he relaxed his ample frame into a leather armchair. 'I'll be damned if I'm going to freeze my nuts off in some river in the North of England.'

The college was not particularly big. It was made up of several buildings, all with interiors that had a nautical theme or resembled the rooms in a superyacht; fine wood panelling, bespoke furniture and computers were all over the place. We settled down to begin the course, watching a presentation on sea survival starring Olympic swimmer Sharron Davies. Some of the trainee deckhands had just been in the pool, learning to scuba dive as part of the 'extras' that the college offered. The large window from our ground-floor classroom looked directly across to the pool, and it was hard to concentrate on the intricacies of personal locator beacons while the athletic stewardesses unzipped their wetsuits and dried off.

These distractions aside, we soon found ourselves in the pool as well. Ed, in his swimming shorts, cannonballed himself into the water, while the rest of us jumped in. It was a big pool, deep enough for the students to learn to dive and wide enough to accommodate a lifeboat. The college director threw in a lifeboat and pulled the handle to inflate it. There was a hissing noise as it unfolded and began to take shape: a four-person boat for the four of us. The next stage was getting in. I remembered the director's warning: 'There's not really a right or wrong way to get in, so long as you get in. But there are easy ways and less easy ways.' I was pretty sure that I could get in without assistance, as were the other two officers, but Ed, being quite a large man, needed some support. He grabbed onto the hull and found a lanyard that he could use to pull himself clear of the water. The three of us, behind his substantial arse, then proceeded to push, with not inconsiderable force, to get him safely aboard.

Following him into the lifeboat, we were soon sat on the rubber floor. Taking hold of some small plastic oars, we began to paddle around the pool. Satisfied that we were able to move around, the director began spraying us with a hosepipe and began a wave machine. 'I'm feeling sick,' laughed one of the officers. He looked a bit sick too. To be fair, the pool was moving vigorously, but as we returned to the classroom after lunch the officer, now recovered, pleaded, 'Don't tell anyone I got seasick in the pool, yeah?'

The following day was just as comfortable. I enjoyed it. I would have enjoyed training in a more realistic environment too, but as far as hardships go, a heated pool is not much of a complaint. Handing our sea survival equipment back, we made our way from the pool house back to the classroom.

'Well, that's the wet part over with,' the college director said as we sat down.

On our desks were sea survival certificates, printed up with the college name and official accreditation. I thought to myself that it was probably the easiest course I have ever done. Given the potentially lethal consequences of the ocean, it seemed all too easy to get qualified in sea survival. But I had to remember that this establishment was not set up to train police firearms officers, but young men and women looking to enter the world of yachting.

'Next we will get you qualified in Radar and VHF radio,' the director announced. 'It's a little drier – both in terms of content and water – but we do have our own vessel where you can practise using the radar. Once you've finished in the simulator, we will see how your sea legs are.'

We looked at our pool-sick companion and he looked at us.

'Sound good?'

It sounded excellent to me.

Along with sea survival, all marine escort officers need to be licensed to use a VHF radio and to read and understand a radar picture. If it was necessary to hail another ship or to send out a message, the only way to do that was by VHF radio. Satellite coverage isn't always reliable at sea. Officers also needed to understand radar and other forms of surveillance as part of their regular duties. Once trained, they could go on to learn how to operate more complex systems, the details of which cannot be described in this book.

Our small yacht, piloted and crewed by the college staff, exited the harbour, and we applied what we had learned in the classroom and simulator to the maritime environment. The craft was around 80 feet long and more like a pleasure boat than a yacht, but it contained all the facilities required to train us in operating the radar. It was all quite simple. Once the radar was calibrated, it was easy enough to match the reflection from a boat on the screen to one out at sea. Picking up speed, the boat began to bounce more, meaning there were more radar reflections as the bow of the yacht and radar antenna pointed down towards the sea; it isn't only metal objects that return a signal, but waves as well. We took it in turns calibrating and learning how to troubleshoot common problems. One problem we were not able to address was the feeling of unease our pool-sick officer was feeling every time he tried to focus on the information screen. No sooner had he looked at the screen than he looked away, out of the bridge window, taking a few deep breaths and repeating the process. It didn't go unnoticed. The college staff exchanged glances with each other as they went about their duties. Ed, too, had noticed.

'Right, you've had enough playing with that thing, let me have a go.'

Stepping back, the officer made his way up a short stairway to the open air, while the boss started to go through the procedure for shutting down and restarting the radar.

'Is he alright?' someone asked.

'Of course he's alright,' Ed replied. 'He just skipped breakfast, that's all. Most important meal of the day. Never skip your breakfast.'

Excuses made, I got the feeling that the newly promoted chief inspector would be having an informal chat with the lad to see how he felt about continuing with the training.

Now able to safely operate aboard the ships, I rejoined the reserve team at Barrow. While the inspectors and chief inspector met with the ships' officers and executives to discuss plans and contingencies, we set about practising our intervention skills. Many of these skills were similar to the emergency entry and AFO skills that we trained in regularly. However, there were new skills that we needed to learn or remind ourselves about as well.

'Happens every time,' Kenny said. 'I get all this gear on and now I'm busting for a shite.'

Our firearms trainers had discussed with the higher authorities and the military what tactics we would be expected to employ during the pre-embarkation training. They needed to be relevant to the tasks we were carrying out. Aside from manning the heavy weapons, the main focus of the escort training was to deny access to the cargo on the ship. We stood as a team in one of the *Heron*'s holds. This huge, hangar-like room was more than adequate to act as a classroom, equipment room, breakroom and maintenance room. Dressed in our 'black kit', we were being briefed by the firearms instructors before breaking off to have briefings with our sergeants. It was the first time that the whole *Heron* team and reserves would be working together, so the idea was to start simple and move to more complex action later.

The officers were buried in police equipment: firearms, pouches, radios, spare magazines. It felt good to be onboard and taking some steps closer to the operation.

'To start with we will go through some gangway scenarios. That means the reserve teams need to be shore side, and the rest of you on the ships,' Jacko announced. 'It's also the most likely time that the ships could have unwanted visitors as well, so we will revisit the scenario regularly and up the ante as the training progresses.'

While in port in the UK, the ship is technically private property, so any trespassers would normally be dealt with under the relevant laws. The fuel that we would be transporting would be loaded in Cherbourg, France, so French laws would apply there, but UK law would still apply aboard the ships. If protesters wanted to get aboard, the easiest action would be to raise the gangway, which was almost like a drawbridge. However, if they somehow managed to get aboard then they would have to be arrested in accordance with UK law. This meant there was a need to revisit general policing skills as well as armed policing skills.

'If protesters do get aboard,' Chris added, 'it means two things. First, that the gangway security team fucked it up and second, that you must not treat the boarders like the those who boarded the *Sevmorput*.'

There were a few laughs. When protesters boarded the *Sevmorput*, a Russian nuclear-powered cargo vessel, the captain ordered his crew to 'assume the worst' despite the less-than-violent actions of the protesters. Using the various hand tools and equipment on board the ship, several of the protesters were badly beaten and thrown into Murmansk harbour.

'So follow the normal arrest procedure. Remember a lot of these protest groups bring cameras and recording devices so you have to be correct in everything you do. Right. Let's get started.'

The first few days of training were not overly taxing. Protesters, drunk visitors and unescorted personnel scenarios formed the bulk of the training. We even had to conduct a ship-wide search to prepare for an investigative journalist managing to sneak aboard undetected while in harbour. But as more variables were added, we knew that it was only a matter of time before we would be training for firearms scenarios.

'QUICKDRAW QUICKDRAW QUICKDRAW' came the announcement over the ship tannoy as we were enjoying a short break. The officers deployed to their predetermined areas on the ship. Some officers went to the bridge, where they could visually identify any threats, along with the captain and his crew. Others ran to the heavy weapon positions that they were expected to crew. The rest of the officers made their way to the deck to cover vulnerable areas and to keep a lookout and receive updates from the control room.

Those officers who were manning the surveillance systems along with the watchkeeper kept a running commentary going of what they had identified with their equipment. We could hear that they had identified a fast-moving contact and were tracking it as it made its way to the ship. The captain and inspectors gave the authorisation to arm the weapons. Bullet belts were loaded and power switches turned on. Officers patrolled the deck, looking in the direction of the last reported contact, while others kept a vigil on the vulnerable areas of the ship in case this was a diversion. The ship's crew had secured the engine room and cargo holds, making them impossible to access. Within minutes the *Heron* had gone from a ship at peace, moored at Barrow-in-Furness, to a floating fortress.

The tannoy sounded, again. 'Endex. End of exercise.' The fast-moving contact on the radar screen had in fact been a simulation – as if the ship was in a much larger dock than Barrow. The scenario, designed to replicate that of the USS *Cole*, was intended to bring the crew to readiness as quickly as possible.

'Too slow. Far too slow,' Chris said to us in the debrief, but in particular to Inspector Mills, who had scrambled from his meeting once the quickdraw was called. He was the tactical commander of all the firearms officers aboard.

'Boss, you need to make sure that once that contact is seen on the screen, someone identifies it. It's no good waiting for the gun crews to tell you what it is. Get in touch with the bridge officers or the ship's crew. Find out what they can positively identify and then use that to inform your next move.'

The inspector nodded. As a sergeant he had been on the MEG for a long time. Now promoted, he usually ran the Sizewell power station policing unit. Tall and lean, he was a natural sailor and one of the few inspectors available for international voyages. Unfortunately, he was not a very competent tactical commander and relied heavily on his sergeants to expedite his decision-making process. The firearms instructors knew this and once even wondered, 'Did he get this job by accident or what?' He was, however, easy to get along with – and on long voyages this is essential.

'If there is a quickdraw, then there is no excuse for hanging around.' Jacko was speaking to the rest of us now. 'There is no time to put your coveralls back on. If you took them off for whatever reason–' he glanced at Jimmy Guns '–then just throw your armour and anti-flash over whatever you're wearing and get to your station.' Jimmy Guns had taken his coveralls off to change his clothes underneath, which had become wet with sweat from the morning's activities. We were all tired and sweaty from navigating the many passages and areas aboard the ship.

'Pirate rig is good enough for a quickdraw,' Jacko concluded. 'Speed is what we want lads. Speed. When FOST comes to test you, you can bet they will be testing you with a stopwatch.'

After endless scenarios and exercises, the entire escort group made its way south to Portland Harbour to conduct some interoperability training on the *Sir Tristram*. After a day of that we were to rejoin the *Heron* and the *Pintail*, which were currently making their way south to take part in sea trials; before setting sail for Japan, they needed to be tested for seaworthiness.

We only had one day left to train, and the mood among the officers was high. Grey and rust-stained, the *Sir Tristram* had been in service since the late 1960s. It had seen action all over the world, including during the Falklands conflict, and was now relegated to static duties, allowing the military and various agencies to practise tactical skills aboard without needing to worry about interfering with the crew or damaging vital equipment.

We weren't going to use the whole ship, however. Being considerably larger and of a different design to what we had been training on, the plan was to conduct training in the accommodation area of the ship. This was new; our ships had their crews already aboard, and bursting into someone's quarters would not have gone down well. This training was also a chance to meet our counterparts in the military who would support us if control of one or both of our vessels was lost.

We were getting kitted up in one of the large cargo areas that had previously been used to shift vast amounts of military hardware: vehicles, troops, tins of beans. None of us had worked alongside the military before in our capacity as police officers, although a lot of the group were veterans. As a result of our meeting with the Royal Marine major, we were first going to display our capability to the team that had joined us aboard. Much like us, they were dressed in black and equipped with various items that they would need to take part in the exercise. Suddenly, the lights went out, plunging us into absolute darkness in the windowless cargo area. The ship was drawing power from 'alongside' and presumably someone had accidently flicked the wrong switch.

'Oh for fuck's sake,' came a voice from the other side of the hold. Clearly an operator from the other team was feeling frustrated about this inconvenience.

'Scared of the dark are you?!' Jimmy Guns laughed. The sound of his soft Cumbrian accent indicated that he was only a few metres away from me and the rest of the reserve team.

When the lights flickered on, there was a steely-faced marine with us as well. Inches away from Jimmy Guns, he slowly shook his head and said something into his ear. I think the general understanding they had reached was that, no, he was not afraid of the dark.

I was surprised to find that the tactics the military used were not too dissimilar to ours. After they had seen our efforts in recovering control of the accommodation block, they took it upon themselves to show us how the true professionals did it. Clearly more experienced and with noticeable speed and aggression, they achieved the same results as us in around half the time. During the debriefing, Ed and the inspectors thanked them for their participation.

'Why don't we just send these guys instead of us?' Craig asked, almost as a rhetorical question. We knew why, but Inspector Mills, who was standing with the other tactical commanders, answered anyway.

'Because it's a civil vessel, Craig. With civil material. If it was a military vessel with military personnel sailing into foreign waters, that would normally be called an invasion.'

He was correct, of course. The political agreements that many nations had agreed regarding civil nuclear materials was that the military would not have access or control over it, unless there were multilateral agreements or dire circumstances. With the wheels of politics moving so slowly, the job of escorting nuclear material internationally, in the UK at least, is the responsibility of the PNTL, the UKAEA and the CNC.

'What did that guy say to you Jimmy?' Danny asked. We were enjoying a well-earned drink in the nearby Harbour Hotel.

'He said,' replied Jimmy, between mouthfuls of his thrice-daily chicken salad, 'I'm not scared of the dark and I'm not impressed by you.'

'Lesson learned there then, Jimmy!'

'Yeah, lesson learned the hard way. It just came out automatically, you know.'

It was important to be able to get along with everyone aboard the ships – especially during those periods where fatigue and a lack of sleep might cause tempers to be shorter than normal. Storms at sea can last a very long time, causing many crew to have broken rest periods. Along with spontaneous and preplanned tasks making shifts longer and rest time shorter, the officers need to also maintain a good sense of humour for the wellbeing of the entire crew. Nothing spreads faster than a bad mood.

'All set for tomorrow then?'

Danny was addressing his entire team this time. We were going to put the pre-embarkation training to the test as part of an exercise. If FOST deemed us unsuitable, the whole operation could be cancelled, causing considerable embarrassment for the CNC and major political headaches. If the firearms instructors determined that an officer wasn't performing well, one of the reserve team might be moved in to take their place. It was as much about individual performance as it was team performance. The crew were aboard as well, and the captain and his team would be running drills concurrently.

We knew that the captain could require us to become involved with medical exercises, flooding and fire exercises as well as man-overboard drills if he so wished. Most of our team had been able to cover this before we started the tactical phase of training. I had not, and had only been able to snatch a few brief conversations with other officers about what might be expected of me – one of the drawbacks of being a last-minute member of the operation. I had been assured that, if I found myself aboard during the operation, one of the crew would teach me everything I needed to know.

The Royal Navy's assessors put both the *Heron* and *Pintail* teams through extensive exercises at sea. Some of these were prearranged and briefed; others were spontaneous. They all required a great deal of teamwork and effort. Fortunately, there was a surplus of officers aboard. The reserve teams would normally guard the gangways then wave the ship off as it slipped its mooring for the high seas. Walking past Ed's cabin, I heard his voice.

'Princess, come in here for a moment.'

I stepped through the half-drawn curtain that separated the cabin from the walkway. The accepted protocol was that if the door to a cabin was shut, the occupant did not want to be disturbed. If there was only a curtain, then it was an open door to any visitors.

'Wanna cuppa?' Ed asked, pouring himself some coffee from the kettle on his desk.

'I'm okay thanks, boss. Do you need me to do something?'

He sat down at his desk. 'I just wanted to make sure you were ok,' he said. 'You looked a little green earlier.'

I was feeling a bit off. Moving through the interior of the ship on rough seas made me feel tired and queasy. I was much better on deck or on the bridge, where I could see the horizon.

'I'm good thanks boss. Just need to get my sea legs.'

He opened his drawer and pulled out a packet of ginger biscuits, then poured a coffee. 'That's going to work wonders. Best cure for seasickness is strong coffee and ginger.'

I didn't know if it was true, but I took the prescription.

'It'll be good to get away from HQ,' he said. 'This new chief bobby, the way the department is being touted as some sort of special forces unit. It's not... You seem to be enjoying yourself though!'

Finishing a mouthful of coffee-soaked ginger biscuit, I said, 'Yeah, it's good. Plenty to see, plenty to learn. Maybe next time I won't be on the reserve team.'

I think that might have been a lie. Being on the operational team would have been an experience but probably a very boring one. Having spoken to the other officers who had undertaken the initial voyage, they made it clear that the reserve team had the best role.

They were able to do all the training and then take on policing tasks alongside the ship while the rest of the crew were aboard. Then, once the ships departed, and assuming nothing happened, it was a long and boring cruise – often through rough seas.

Returning to port after the sea trials, FOST let us know that they were satisfied with the performance and would be giving a full debrief to the military and the upper echelons of our constabulary at a later date. The 'hot' debrief, which occurred almost immediately after we had concluded the training, only identified a few minor issues in command and control. Tactically, the MEG was deemed to be proficient and competent to carry out its primary purpose. All that remained was two weeks of leave for most of the officers before a return to Barrow-in-Furness to begin the operation.

I was taking a long weekend and then returning to HQ to help tie up any loose ends from the PAR and put some finishing touches to the operational order. There was still the matter of a road escort to plan, but Bones had been taking care of that. I reminded myself that I was not a member of the MEG but part of the planning team. Although I had been through most of the initial training, my function was to assist in planning; this part of the operation was merely a secondment for me. I made a mental note to make sure that I would look for some tasks that I could finish for Bones and Matt before I returned to the ships as well.

# 18

I was surprised to find someone else's coffee mug on my desk and a police jacket on my chair. '138' was printed on the epaulette. I didn't know who PC 138 was, but I was definitely going to get my desk back from them; it was in the corner of the office, so it offered the greatest privacy from any chief officers or staff that might wander into the room looking for volunteers. It sat on the opposite side to the boss, Ed, as well, meaning that Bones or Matt would normally be in the firing line for tasks before me. Being the lowest rank, however, meant that such things were usually delegated downwards to me anyway.

Once I had taken my seat and cleared my emails, Superintendent Watson walked in accompanied by two men in suits. They were around the same age as him, in their early fifties, but were in better condition. The superintendent was not the most athletic officer in the constabulary. Perhaps it had something to do with his role; he had no time for fitness, and only ate convenience food.

'You were at Oldbury before this weren't you, Okuhara?'

I had stood up as soon as they entered the room and was about to suggest that they look for one of the more senior members of the team in order to help them out.

'Yes, sir.'

He nodded and turned to his two companions. 'This young constable will drive you to Oldbury and back, he'll pick you up in around twenty minutes' time. Why not have a coffee whilst I get him all sorted out?'

They smiled and nodded at me, and said to the boss in a foreign accent, 'Thank you, Wally.'

I had not been expecting any more chauffeur work, having just got back to HQ. It was becoming a regular task for me. But I thought that a day back at Oscar Yankee might be nicer than clearing emails. Taking a seat in Matt's chair opposite me, the superintendent told me about his two companions, who had just left.

'These are personal guests of the chief constable, from when he was in the diplomatic world.' He put a strain on the word 'diplomatic' and made some quotation marks with his fingers. 'These two men from Sweden are senior officers in the National Task Force, which is their counter-terrorism unit. Yesterday they visited Harwell, they saw the HQ, now I need you to drive them to the closest power station unit, which is Oldbury… how long will it take if you leave now?'

I knew nothing about the National Task Force, and little about Sweden, so at least I would be able to talk to them about something, provided they were in a talkative mood during the two-hour drive.

'A couple of hours, sir. Probably a little longer on the way back with all the commuters on the M4.'

Apparently the chief was keen on showing these friends in high places his new command, having been in place for a little over a year now. He spent most of his time in London and was not often at the HQ, preferring to use his influence in the 'corridors of power' in and around Parliament, Whitehall and Thames House. Matt had told me that although it meant a lot of time away from home as his driver, it also meant a lot of time doing airport runs and overnighting in the London clubs where the chief preferred to accommodate himself while away from home. I liked the sound of that, but the best I could manage, for today at least, was a trip to a nuclear power station. It would be good to catch up with the crew at Oldbury, though, and I hoped that the officers on duty were some of my previous shift mates.

On the way down, our guests explained to me that their role in the National Task Force for the Swedish Police Authority didn't just cover counter-terrorism work but also international work. They didn't say much about their relationship with Mr Thompson, other than they were all friends; and they didn't suggest that they were going to work with the CNC at some point as part of their international remit. In fact, they had quickly changed the subject to how I felt about the latest parliamentary scandal, in which Members of Parliament had been taking advantage of the expenses system and profiting quite nicely from the taxpayer.

I knew full well that some officers in the CNC were occasionally overly generous to themselves when claiming expenses, but I had never been one of them. During the previous month and a half of training, I had seen officers allowing themselves a drink or two plus a three-course meal every evening, as well as officers who would make do with an orange juice and a salad. It was all well within the rules, but the MPs in Westminster had really been taking the piss with property expenses that included mortgage payments and salaries for staff that weren't actually carrying out any work. I suppose that the senior officers I was driving didn't want to discuss policing with the driver – as by the time we had arrived at Oldbury, the state of British politics had been thoroughly debriefed.

When we arrived, I found that Mo was still running the policing unit, and Beegee was his deputy. Mo brought the Swedish delegation into his office while I headed straight to the kitchen for the spare mugs to get a cup of tea going. My old shift wasn't on duty; it was B shift, made up of Gav, Fish and two new faces. Mal was a young and energetic kind of officer – although he eyed me with some suspicion as I made my way into the crew room. He and Gav had been waiting for us so that they could guide the visitors around the site – much like Dave had done for me on my first few shifts. They were to get a full tour, excluding the reactor areas.

'Who are they, then?' Mal asked. 'We got told they were Swedish police officials who were going to compare our security arrangements to theirs.'

That was more than I had been told, but I indicated that this was the case and that they were from an elite unit that also dealt with international policing concerns.

Gav introduced me to his new probationer and explained that the station had taken on several new officers. Predictably, given the CNC's high attrition rate, two had already left.

'You know how it is,' he sighed. 'It's the boredom. Some people can't hack it. Some people use this as a stepping stone to the home office and some people got sold the wrong idea and go back to what they were doing previously. Veterans tend to love it, uni graduates not so much.' He playfully punched Mal, who had been studying archaeology before coming to the CNC. There were many occasions on shift where nothing happens at all – and then very occasionally there is some police work to do.

'You heard from Dave?' I asked.

'Oh yeah, he popped in a few weeks ago. He was on his bike. Loving it in West Merica by all accounts.'

That was good to hear. Whereas the CNC suited Gav, Fish and me, Dave was not really in his element and needed something more. He had joked before leaving, 'No more gate guarding! Time for some real policing.'

'So has much been going on?' I took a seat at the crew room table. I had sat at this table countless times, and didn't expect to be sat here again quite so soon. I felt nostalgic as thoughts of grabbing a quick bite in between patrols came back to me, or doing some paperwork because there was no room in the office. I hadn't been in touch with the team at Oldbury, or checked any reports for the southern BCU. I had been too busy with Escort Ops. Gav took a seat too.

'We've had a few things. We caught some poachers out on the river. Poaching for eels!'

I had no idea eels could be illegally caught. I knew that fishing required a licence but had never really thought much about it.

'Oh yeah? What happened?' I asked.

'It was just by the south lagoon, one evening. I saw a car that I didn't recognise so me and Mal went over to have a chat. Once we saw all the gear and buckets he had, I thought that it seemed a little overkill for a spot of fishing.'

I didn't realise Gav knew so much about angling. Given that he was almost always on overtime, I assumed his hobby was coming into work.

'I took a bit of advice from the Environment Agency and nicked him. Avon and Somerset were pretty confused about it too, but the agency sent down a water bailiff to help prepare the case.'

I couldn't think of a policing scenario further away from nuclear security than poaching. But seeing as the poacher was on the south lagoon, that placed him well inside of the CNC's jurisdiction.

'And you heard about the murder, of course.'

I had heard. A road maintenance worker found a binbag containing bones just outside of Thornbury, close to the power station. They belonged to a person who had been missing since 1996.

'Avon and Somerset were basically left with no coppers in Thornbury whilst they closed that section of the road down, so we had to backfill for a couple of hours.'

I hadn't heard about that.

'What you had to do real police work?!' I feigned astonishment.

'Nah, not really. We just had to provide some high-visibility patrols until the real police got back from guarding the scene. Hey, they should have asked us! We're great at guarding things.'

We talked some more about what was changing, about the proposed shutdown of the power station and about Escort Ops when Mo finally came into the crew room.

'You ready Gav? Mal? Are you coming as well, Matt?'

Rather than sit around doing nothing, I decided to join the power station tour, although I let Gav and Mal handle the particulars. It felt good to be walking around the site again. Not a lot had changed. The guards were the same, the health physics guys were the same and everyone was going about their work just like they had a couple of years ago. The well-established protocols of nuclear generation didn't really allow much in the way of flexibility.

With the tour concluded, my next task was to drive the two delegates into the village where they would meet the superintendent, his secretary Amber, and Mo and Beegee for lunch. There were two pubs in the village, and as I had learned during my time at the OPU, one pub was for visitors and the other pub was for locals. I was expecting to wait in the car while they ate but was surprised to be allowed in to join them.

'It's not dissimilar to home,' one of the Swedish delegates said. 'Semi-rural. Quiet. Easy to police.'

Pointing at the old oak beams in the ceiling of the pub, the Swedish official queried, 'Is it an old settlement?'

This caught Beegee's attention. Despite his London origins, Beegee had fallen in love with the area and was becoming fully engrossed in village life. He had even joined the parish church council and helped to plan the annual summer fete.

'Very old,' Beegee said, jumping in enthusiastically. 'Very old and very interesting.'

The visitors were no doubt trying to make polite conversation but had inadvertently opened a door to the deputy's latest interest.

'This area is full of ancient facts and legends.'

I didn't know much about legends in Oldbury, and I had grown up nearby. A whale had beached itself on the riverbank in the 1880s, I knew that much. There was also a ship graveyard downstream, where the wooden hulks of long-abandoned vessels were still visible. I didn't know much more, other than that people would regularly surf on the tidal bore in the Severn.

'You mean the whale over at Littleton?' I asked.

He shook his head. 'No, not that – the Cavaliers lost in the Severn and that ferryman.'

This was news to everybody. Between mouthfuls of pub food, he elaborated.

'Back in the days of the civil war, this ferryman at Aust worked the Severn, moving folk across the water. Parliament's man. But when two Cavaliers, Royalist messengers, needed passage to Wales, he agreed.'

During the English Civil War in the 1640s the country had divided itself into two factions: those who favoured an elected government, known as Parliamentarians or Roundheads, and the Royalists, known as Cavaliers.

'Here's the thing. He was sly. Instead of a safe ride, he left them stranded on a sandbar, halfway across. Rising tide swept them away, their screams drowned by the water. They say, even now, those screams echo along the Severn's banks. Some swear those ghosts still fight the current in the river.'

It was the first time I had heard that legend. It sounded plausible enough. Gloucestershire and Bristol saw their fair share of conflict during the civil war. But I had done scores of night shifts along the river and never seen or heard anything supernatural.

Beegee leaned in. 'During one of our night patrols, Mal stopped dead. "You hear that?" he said. I thought I heard something too. An animal, maybe an owl, but sounding deeper, eerie.'

Even if it was just a legend, he was spinning us a great yarn. It made the long drive from HQ worth it. A catch-up with the lads, a free meal and now a legend about civil war ghosts.

'Could be wildlife,' Beegee continued, 'but doubt seeped in. Was it just nature's tune or something older, darker? So I asked around. I heard from the parish council and the vicar himself about the traitorous ferryman and those lost souls.'

Returning to HQ, I found that the PC138 who had taken over my desk was assigned to Escort Ops to act as a welfare officer. After I had taken my place in the team, it had become apparent that we lacked any welfare provision to manage the needs and concerns of the officers. Things like childcare, short-notice family emergencies and a regular point of contact for the families of deployed officers had not been included in the PAR. Fern, who had been assigned to the superintendent's office, had come across for a few months to carry out this new role, which only needed to be temporary. She had been in the police for most of her life. Starting out as a special constable in the Met in the 1980s, she had moved to Oxfordshire with her husband and joined the UKAEAC at Harwell; she had been there ever since, and was one of only a handful of PCs who were not firearms officers.

Fern had a list of requests from the officers who were going to set sail in the near future – everything from football results to birthdays and weekly horoscopes. She also had information about what to do in case of a bereavement. Did the officer want to know immediately, or find out after arriving back in the UK? Once at sea it would be impossible to get the officer back, and the stress of such a loss could prove detrimental to the entire ship's security arrangements. Information had to be transmitted by satellite to the ships when a signal was available. In some parts of the ocean, particularly close to territorial waters, the signal was almost constant. Otherwise, most data was sent in a burst and downloaded when possible.

Fern had also been taking on the non-confidential matters in the PAR, such as arranging passes for dockside visitors, hire cars and bookings for conference facilities and the all-important food arrangements. Normally Candice and Amy would handle this sort of thing, but with the department running two concurrent operations it was useful to bring in an extra person. Anything that was deemed classified or secret, however, had to be undertaken by one of the officers with security clearance.

Less than two weeks out from Operation Accord, almost everything was ready. I checked my items in the PAR and found that very few still needed to be completed. There were a few kit and equipment requests to sort out – things that had been brought to my attention while training – and I still needed to print off and distribute several copies of the operational order that had been produced by Matt. He had finished it weeks ago but was updating and adjusting it as needed. Now, with the deployment just around the corner, he asked me to get the copies sent. It was an easy job to do, but it was very boring. I just sat by an industrial-sized printer, taking the pages as they came out to make sure that none of them fell into the wrong hands. Not that espionage has been a problem, to my knowledge at least, in the CNC HQ. Forming the operational order into neat catalogues, I thought ahead to the road escort job that would happen shortly after we got back from France. I would have to find Bones and have a good catch-up with him to see if he needed anything done, and to also find out my part in the operation.

'Control tower officer,' Bones said to me, swinging his computer screen so that I could see it better. 'You're going to be the control tower officer.'

I scanned my eyes up and down the list, checking the officers numbers against the assigned roles.

'We've got twelve officers from BCU Scotland to act as perimeter security at the airport as well... It's a big job.'

This was a first for the CNC. Previously all parts of Escort Ops had been handled by officers specifically trained for the role. However, all the CNC's police officers held the office of constable – or held full police powers – when at a trans-shipment site. In this operation, Wick Airport was the trans-shipment site and additional security was being provided in the form of off-duty officers at the Scottish nuclear power stations.

'What's the score with being a control tower officer?' I asked. I was familiar with all the roles on the spreadsheet – reconnaissance, support, driver and such – but I had not heard that role before.

'It means you do naff all and sit in the control tower to make sure nobody interferes with the air-traffic controller. The Americans specifically requested that, so you're the one doing it. It won't require any additional skills, and with your marine escort training, you're more qualified than the rest of the resources at the airport.'

I had visions of myself sitting on a foldable chair at the bottom of a control tower reading a newspaper, then asking for a password before opening the door to let people in. Bones could see the look on my face.

'Cheer up. You could be putting the whole thing together, like me. Speaking of which, I emailed you a list of things to do.'

I returned to my desk and searched my inbox for my tasks. I felt a little annoyed at being given such a basic assignment, but Bones was right. If this operation had come at any other time, I would not even need to travel to Wick. Instead I would be tapping away on a computer, making phone calls and drinking coffee. At least this way I had a role, albeit a minor one in the deployed part of the operation. It was a large multi-agency job as well, so there was always the chance that things would change.

My phone rang.

'Escort Operations, PC Okuhara.'

No answer.

'Hello, Escort Ops here.'

No answer.

Assuming that it was a wrong number, I hung up the phone, only for it to ring again almost immediately.

'Escort Operations, PC Okuhara.'

'This is Inspector Thorpe from Sellafield,' said an impatient voice. 'If I don't speak when I am on the line it means I am not free to talk.'

I thought that this had to be a joke or wind up. 'Come again?' I replied.

Sterner words this time. 'If I do not speak when I am on the line it means I am not able to. Do you not hold security clearance yourself?'

I was beginning to suspect that this was not a wind-up, but it still seemed pretty odd to me. 'I appreciate that, sir, but you called me and I'm not a mind reader so... Is there anything I can do for you?'

'Yes, I need to know the dates for Accord so that I can arrange overtime.' The sentence came out almost as one word.

'You'll need to speak to your planners I'm afraid. The dates are still coded.'[1]

*Click*. The caller had hung up.

Bones was looking across at me. 'What's up?'

I explained to him what just happened.

1 Coded dates are those that have been encrypted as a numerical code so that the information cannot be accessed without a key.

'Thorpe? Oh yeah, I know him from when I was at Sellafield a few years ago. He's weird. He was a sergeant at the time but pretty unpopular. Always stressing. Fucking idiot.'

I had heard that phrase come from Bones more than once. I hoped that I wouldn't fall into that category, but it seemed at some point everyone did.

'Yeah, can't give him the dates though. Email him back and copy me in, I'll sort it before it grows arms and legs. He'll try to go over or around us to get what he wants, if he's anything like he was before.'

Finishing my work for the day, I went to my girlfriend's place to pick her up and enjoy an evening out. It went well; we found out we had in common than we had first thought. But she had to get up early to help with her host family's morning chores, so I took her home, then made my way back to Cheltenham. While stopped at some traffic lights, the driver-side door of the car ahead of me was suddenly flung open and the driver emerged and ran to the car in front of him. He opened the other car's door and threw a heavy punch before returning to his vehicle and driving off through a red light. As I was about to get out and check on the occupant of the other car, it began to move as the lights changed. Only 50 metres down the road, however, the hazard lights came on and it pulled over. I did the same, then got out and went to the passenger window. The young lady driving the car was upset, and understandably so. She was developing a red mark on her cheek and had clearly been crying.

'I saw what happened. Are you OK?'

She sniffed and nodded.

'Do you want me to call the police?'

Again she nodded, and I called Glospol. I explained to them what I had seen and told them I would wait with the lady until they arrived.

'Is there anyone else you want me to call?' I asked.

'My son,' she said. 'Tell him mummy will be late home.' She handed me her phone which was already ringing. A young voice answered, surprised that it wasn't his mother that was calling. He was with his grandparents and not alone, but I did as she asked and told him, 'Mummy is a bit busy helping the police so she will be home a little later than usual.'

Glospol arrived and began looking into the incident. They asked me what I had seen and gave them everything I could. I had noted the number plate before he drove off as well as a description of the guy.

'Can you come to the station and give us a statement?' asked the Glospol officer. I said that I would.

I heard the young woman explain that she had just broken up with the guy, and that he had not taken it well. Not long after she had left his flat, he decided to chase after her car, ending in the assault. He was going to be easy enough for Glospol to pick up; they knew everything about him already.

I sat down at the desk in an office at Gloucester Police Station and the PC began to take a statement. Name, age, address, occupation...'

'Police officer.'

He looked up. 'You in the job?'

'Kind of. I'm in the Civil Nuclear Constabulary.'

He clicked his pen and wrote it down. After the statement was complete he asked me if there were any dates when I would not be available if this went to court.

'Sure, any time between next week and the next four months,' I told him.

'Going travelling are you?'

'Something like that.'

# 19

The initial phase of the operation – the preparation for securing and loading of the fuel – was due to be carried out over the course of several days in both the UK and France. The entire escort group had a briefing the day before the operation was due to start. Afterwards, the senior officers held a meeting in the Abbey House hotel, where the reserve officers were also staying. No Princess Suite for me this time. I took notes during the briefing and then helped Ed put his tactical briefing together for the officers on the ships. The senior officers said their pieces before outlining some of the particulars. We had officers from Special Branch, ONR, Cumbria and Northumbria with us as well. Many of these people would not be dockside for the operation but still needed a solid understanding of what was going to happen.

Arriving at PNTL's terminal at Barrow-in-Furness, the reserve crews were stationed at the gangways to their ships and inside the dock area itself. The briefing for the dockside part of the operation had happened earlier that morning, while it was still dark, and involved units from several different constabularies. Cumbria Constabulary had provided a public order unit and stationed it nearby, but away from the dock – which was behind a substantial amount of physical protection, with fences and gates limiting access to authorised personnel only. There were also plainclothes officers patrolling likely meeting points for any preplanned protests that the CNC's Special Branch detectives had identified beforehand. Northumbria Police provided divers and a fast boat to check the ships before they set sail as well monitoring the waterways leading out into the Walney Channel. Finally, the CNC provided armed protection aboard the vessels and dockside policing capability. It was a huge operation, but we were hoping it would go unnoticed by the public.

It was a cold and wet morning as we began our duties just after sunrise – ideal conditions for what we had been planning for the past several months. I was standing at the top of the gangway with Jimmy Guns while the rest of the reserve team patrolled the dockside and checked credentials before allowing anyone onboard. Wearing our flame-retardant black coveralls, we carried general policing equipment: batons, CS spray and handcuffs. We each carried two radios as well, one to communicate with the ship and the other to speak to the officers patrolling the trans-shipment point and the command-and-control centre. While we reserve officers were taking care of gangway security, the rest of the escort group were aboard, manning the surveillance systems and patrolling the deck. At this point, the ships were at their most vulnerable.

Still drawing power from the dockside, the crews were going through their final preparations before starting the huge engines and setting sail for their first port of call: Cherbourg. It was first thing in the morning and the sun had been up for less than an hour. I felt quite chilly when the wind blew in from the Irish Sea, so I tried to position myself out of the breeze.

'Who's that?' I asked Jimmy, pointing my finger to a moped travelling outside the terminal perimeter. 'He's got to be riding on the grass; there's no road over there. What's he doing?'

Jimmy picked up some binoculars to have a better look. Being around 100 metres away, but with several obstructions between the moped and us, it was hard to get a decent look.

'You reckon he's a nerd or something? Looks like he's carrying a camera or optics... Can't make a lot out. Helmeted. Full-face silver bike helmet. Quite a tall guy. What do you reckon?'

As this was a trans-shipment point, the CNC had no police powers outside of the terminal; although going and speaking to someone is not necessarily an activity that needs police powers. There were no reported threats to the ships, but if this person knew that they would be loading and sailing – at this precise time, before most people are even out of bed – it could indicate a compromise in the timings for Operation Accord.

'Better let Cumbria have a chat with him,' I concluded.

Jimmy passed me the binoculars to confirm everything that he had said to me and contacted the operational commander.

Nearby, in the old Barrow-in-Furness Borough police station, the rest of the planning team had set up a command-and-control centre. The old police station, built just after the Second World War, was not an ideal command post. The space in the old brick building was limited, and moving in all the equipment needed to run an operation of this size made for a tight squeeze. It was close to the ships, though, meaning officers could come and go from the police station to the dockside with ease.

Bones, manning the communications, took our report and passed it to the superintendent, who was in overall charge. Along with Matt and some officers from Cumbria Constabulary, the command team were warm and comfortable compared to the officers on the dockside. Listening into the shared radio frequency for the operation, I heard an unfamiliar voice call for one of the nearby Cumbria callsigns. We knew from the briefing that he was requesting that an unmarked vehicle travel to the end of the road that led to Cavendish Dock, where we were stationed.

'CNC dockside are reporting a moped, red, with a male rider, silver full-face helmet, black outdoor-style jacket, brown cargo pants, so far?'

The response came immediately. 'So far.'

Giving the location, he added, 'Can you head over and find out what he's doing? They suggest he might have a camera or optics and is showing some interest in what's going on. Try to find out how he knew to be here, at this time, over.'

They seemed like very specific instructions. I guessed that our Special Branch officer had something to do with that. Acknowledging the request, the callsign made their way to the reported location, estimating it would take them ten minutes from their current location on the far side of the dock.

'I bet they thought it was going to be some quiet overtime,' Jimmy said. 'It probably will be. It's just a lad on a scooter... But still... We better head on down and see what the craic is as well.'

I agreed. Ten minutes is a long time in policing. Using the VHF radios that communicate with the ships, we let the officers patrolling the decks know what we had reported and what was happening. Looking up, I saw one of the balaclava-clad escort officers step out onto the bridge to train some powerful lenses in his direction. I saw one of the several external cameras, controlled from within the ship, turn in that direction as well. High up on one of the masts, both the *Heron* and the *Pintail* boasted an impressive array of communication and surveillance equipment. The moped rider probably didn't realise just how much attention he was drawing to himself, either by accident or by design. I made a short transmission to the *Heron*. 'We're going to head over and speak to him.'

We left the secure part of the dock using a gateway almost identical to those found at nuclear power stations and headed down the road and into a car park used by the dock workers. We found the moped rider there. I could see that there were tire marks on the grass where he had been, and mud was stuck in the treads of the moped tyres. He had removed his helmet and was looking through pictures on his camera. Being so engrossed in what he was looking at, he didn't see us coming.

'Alright bud, what you got there?' It wasn't a subtle introduction.

The rider quickly looked up in surprise. That soon changed, and almost immediately he switched off his camera, put it into its black nylon case and secured the whole package under the seat of his moped.

'It's a camera. Why, what did you think it was?' His tone was impatient, arrogant almost. He looked older than either me or Jimmy; I guessed he was close to forty. He was tall and thin, and he looked like he didn't sleep very much.

'My guess is that it was a camera,' came Jimmy's reply.

'Well done. You a detective are you?'

Jimmy didn't answer that. 'What you been taking pictures of?'

'What's it to you?' He'd been expecting that question, no doubt.

'Yeah normally it's nothing, but today it is quite a lot to me, so what have you been taking pictures of, bud?'

He didn't say. Instead he went on about not needing to explain himself to us and about freedoms and rights. Jimmy, clearly getting annoyed, changed his radio channel away from the operational frequency to the county channel; so that he could run a PNC check on the moped, leaving me to try to start a dialogue.

I stood beside him, almost shoulder to shoulder, and pointed my finger towards the *Heron*'s mast. The white tower was easily visible, along with the array of radars, sensors and cameras that covered it.

'That ship has been watching you ever since you came down the road on your moped,' I told him. 'It's watching us right now.'

His eyes followed my finger. I had his attention.

'So we know what you were taking pictures of and when, and my guess is that at the same time the ship was taking pictures of you.' I was making an educated assumption on that point. 'Look, we're not accusing you of anything illegal, but the people on that ship want to know why you are taking pictures of them. That's it. It doesn't need to be an argument or debate about freedoms. So let's keep it simple. What's your name, please.'

He took a few seconds to reply. 'Mike.'

I took that at face value, but if he said his surname was Hunt, then that would definitely be a wind up. 'I'm Matt,' I replied, tapping at the numbers on my shoulders at the same time.

Jimmy was speaking on the radio, confirming the vehicle details with Cumbria before changing to his VHF set and passing the details to the ship. I could hear his voice on the radio that was plugged into my earpiece. The vehicle was not registered to someone called Michael or Mike. I didn't wait to hear the full address, but it was close enough to call him a local.

'So, Mike, is there any reason for you to be taking pictures of the ships today?'

'Yeah, but it's none of your business.'

'Okay, so is there any reason you are riding someone else's moped today?' I mentioned the registered owner's name. His expression changed and he shifted his gaze from the *Heron*'s masts and towards me.

I gave him a few seconds before saying, 'So is it Mike?'

He was looking more uncomfortable now, but remained defiant. 'Like I said.'

'Look at it from our point of view, Mike. You're taking pictures of those ships, pictures of the officers onboard and you made a hell of a mess on the port authorities' grass when you rode over it.' I pointed behind me with my thumb. I wasn't interested in damage to the lawn, but I wanted to reinforce that there was at least a mild amount of interest in his photography. 'It isn't exactly normal activity you would find on a dockside on a cold wet Sunday morning. On top of that, I think you gave me a fake name. Do you have any ID?' Maybe he did, but he didn't produce any and I had no right to demand any or do anything about it. The trans-shipment point ended at the security gate, so Jimmy and I were just talking.

The radio in my ear spoke again. It was the *Heron*. They had passed the name of the photographer to our Special Branch officer attached to the operation. She found that he was part of a well-known protest group, and had actively been protesting against Barrow Docks being used as an import site for nuclear fuel as well as protesting the existence of Sellafield. Given such protesters have a habit of photographing officers and uploading their pictures to the internet, a lot of CNC and ship personnel wore balaclavas while on duty at the dockside. It was likely that Operation Accord would soon be getting a press release that had not been arranged by us.

Our Special Branch detective wanted to come from the command post to the dockside and speak to 'Mike' herself. The Cumbria callsign that was coming to assist us had to stop to pick her up.

Next, Bones came onto the net. 'Can you try to keep him there? Within reason. Michaela is coming to the dockside with the Cumbrian patrol.'

I gave a short reply: 'Received.'

Jimmy walked back towards us. He was ending a call to Cumbria again.

'Bud!' He called.

Mike turned to face him.

'Is that your scooter, Mike? That's not your bike bud, so what's going on? It belongs to someone else. Did he say you can borrow it?' He tapped the police badge on his chest. 'Does the owner know you are riding his bike around Barrow Docks?'

'Oh for fuck's sake, alright!'

We both looked at him.

'It's not like you need my name anyway. I know the law. You aren't even real coppers.'

Jimmy laughed. 'Alright, Mike. The real coppers will be here soon enough.'

Leaving the moped rider in the capable hands of our Cumbrian counterparts and Michaela from Special Branch, we made our way back to the ship. The final preparations were nearing completion.

Soon we would need to jump into our cars and drive to Cherbourg, where we would once again join the ships – this time as part of a multinational security effort.

'You ever been to France?' I asked.

'Oh aye,' Jimmy replied. 'Went on a battlefield tour when I was in the mob. You?'

We were walking up the gangway to swap with some of the other reserve officers and patrol the dockside.

'Yeah, a few times. Never Cherbourg though.'

As we chatted, the Cumbrian officers that had been sent to the moped rider reported their finding to the command-and-control centre. 'He claims to be an independent journalist and a member of an environmental group,' came the conclusion. 'We've sent him on his way. No further action.'

'Well, that's that I suppose,' Jimmy shrugged.

Special Branch continued to look into the encounter while we made our way to Cherbourg. We were updated about Mike's activities by one of the CNC's detectives, based at Sellafield. The protest group had published an article, complete with images, within hours of us speaking to him:

> Sellafield is planning to ship a cargo of dangerous plutonium under armed escort to France in the next few days. The cargo of plutonium, in the form of weapons-useable dioxide powder – a highly prized terrorist material – will be secretly transported. Carrying security crews from the Civil Nuclear Constabulary, the nuclear gunboats must negotiate the notoriously busy English Channel shipping lanes. This clandestine shipment of highly toxic plutonium puts European waters and communities at significant and unnecessary risk. This could be the first of many plutonium-swap shipments as a number of SMP orders have had to be subcontracted to France and Belgium. A report by the Royal Society on the security of nuclear materials,

published just this week, points to the risks of 'suicide terrorism' and the dangers of nuclear transports being subject of sabotage and their cargoes acquired for improvised nuclear weapons or dirty bombs. A spokesman added, 'The shipment from Sellafield will contain enough plutonium to make a large number of such devices and it is plain stupid that, given this and other warnings in recent years, the government and industry are still prepared to put us at risk with their cavalier attitude to security and public safety.' SMP opened in 2002 and was designed to produce 120 tonnes of MOX fuel per year. None was produced in the first 2 years of operation and to date – after 6 years – a total of just over 5 tonnes of fuel have been produced.

Taking a normal cross-Channel ferry from Portsmouth, we arrived in the French city of Caen. We were able to bring our personal effects and some of our police equipment, but not all, for customs reasons. The French border guards and gendarmerie are not known for being flexible. We would have to make do with what we had. As we were outside the reach of the UK digital network, we had also brought VHF radios that we could use while on duty.

An equipment oversight had emerged while we were in transit as well: we did not know how the French or UK customs services would react to vast quantities of police equipment being driven across the border. Bones was liaising with the embassy in Paris, trying to secure a diplomat to meet us and talk our way through any problems that might arise. As I was allocated most of the equipment issues on the PAR, it was probably my fault for not identifying this issue sooner. I had considered, however, having at least one French speaker in our convoy of vehicles to handle routine matters such as paying for fuel or asking for directions if we ended up getting lost. That was me. Far from fluent, I had been learning the language in my spare time and was impressed at how much I had remembered from my school days.

Kev was following the satnav, which showed a pretty long journey to get to Cherbourg harbour. I was glad that he volunteered to drive, as I didn't fancy driving on the opposite side of the road.

'Let's fuel up now. Hopefully that will be enough petrol to see us through back to Portsmouth.'

I had already proven my French ability, in the eyes of the team, by ordering omelette-du-fromage on the crossing, and I was about to be further tested at the petrol station.

After purchasing some unfamiliar-looking snacks and coffees, we were soon making our way through northern France. In the back, Jimmy Guns was firmly wedged between kitbags that contained everything we needed for the next few days. Sound asleep, he would be working overnight with some of the other officers following in the cars behind.

Most of the security was being handled by the Maritime Gendarmerie, who had a significant number of officers based in Cherbourg. They would be providing overt armed protection and patrolling the waterways in their high-speed boats. The National Gendarmerie, their land-based counterparts, had escorted the fuel flasks to a secret location outside of the city and were waiting to be called in when the *Heron* was ready to start loading.

Our role was the same as it had been in the UK except this time our powers only extended to the ship. Gangway security was to be carried out still, but when dockside the patrols were going to be joint patrols with French officers. Essentially it meant leaving most security matters to the Gendarmes. The guns on the *Heron* were covered, as requested by the French government, and the *Pintail* was due to loiter in the English Channel, waiting for her sister ship to rejoin her. So far so good. Loading was expected to happen very quickly, with each fuel flask being hoisted, then lowered into the huge cargo bay of the vessel. However, the ships were not scheduled to arrive until the next day. That meant that the night-shift officers would be resting up in the hotel while those of us on the day shift would join our French colleagues dockside as soon as we arrived.

I was joined by a gendarme *sous-officier* – or corporal, as far as I could tell – called Herve. He was a young guy, probably around my age or similar, and fit the French stereotype very adequately. Along with his fatigues he wore his dark beret and seemed to have forgotten to shave that morning – or had decided to go with the 'designer stubble' look.

'Bonjour,' I said to him, trying to include a satisfactory accent with my greeting.

He took his hand away from the pistol grip of his rifle and shook mine. 'Bonjour,' he replied, followed by an introduction and a lot more rapid-fire French, most of which I did not understand. I got the gist, though, and soon enough he took me to their dockside command post, a purpose-built truck complete with kitchen, conference table and several communications operators manning radios and cameras. The CNC also used to have a command vehicle, but as it was only ever used once, and only for an exercise, this converted campervan was later sold to the MOD. This Maritime Gendarmerie vehicle, on the other hand, looked like it got a lot of use and the people working in it were effortlessly attending to their tasks and duties. Inside it smelt nicely of coffee. Sat at the conference table were a couple of more senior officers, who were having a conversation and making notes. I could hear the radio chatter in the background but did not understand a word of it.

With a slight nod from the seated officers, and not a word of greeting, I was handed several laminated lists. One had the details of who was expected to come aboard the ship and at what time. Another listed those allowed aboard whenever they wanted, namely those charged with detecting radiation plus the several high-ranking officers that the French authorities had sent to oversee the operation. Given the secretive nature of our deployment from the UK, there was a real contrast in how the French wanted to carry out their phase of the operation. The overt armed security, closed roads and local cops on patrol all around the harbour and beyond was in no way intended to be subtle. They had taken out injunctions against likely protest groups, and had already released press statements advising people to not interfere with the operation.

I handed over the lists to Danny and the other reserve team sergeants, who were getting kitted up from the backs of their cars along with the rest of the shift. Half of the team had gone to the nearby hotel and would return after sunset to take over from us.

'I have no idea where we should meet up with the French and what we should do,' Danny admitted. 'Technically we only have a job when the ship is here. The rest of the time, we're just spectators.'

We knew the reasons for us arriving early were twofold. First, if the ship for some reason decided to speed up, it could arrive sooner. Secondly there was the political element, where all three nations – Japan, France and the UK – needed to be seen taking an active part in the first fuel movement in many years. If things went well, it would be the first in a series of operations; if they didn't, it would be a major embarrassment on the world stage. Even so, we were redundant until the ships arrived.

'I ain't prepared to hang around and do nothing all day,' Danny decided. 'We can't speak French and there's no Brits here so the joint patrols is a nonstarter – except maybe for you, Princess. Ask the French where we can conduct our own patrols. We'll form two groups, got it?'

I made my way back to Herve, who was not too far from us, and explained to him what Danny had said. He nodded and spoke into his radio. Once again we made our way to the command vehicle and once again I stepped inside. A printer was running off some maps of the dockside, and one of the senior officers leaned back in his swivel chair to take the papers. He took a thick red marker pen and started drawing circles in various places. Silent as always, he handed me the paperwork. The marked areas were where we could operate. No doubt they too were unsure of our function while the ships were still so far away. To be honest, I was wondering the same. I guessed that the locations he gave me would keep us out of the way of their security function.

Off the coast of Cherbourg, the *Heron* was in position, waiting to begin its move into the harbour. A French pilot had gone aboard to assist the captain in manoeuvring the ship. As it started to come in, the fuel convoy would also move, ideally meaning that they would reach the dockside at around the same time. As was usual on the initial part of the voyage, several officers were still to find their sea legs and were feeling unwell thanks to the swells of the Irish Sea. If any officers didn't want to continue, this would be their last chance to swap out with one of the reserve team. That, of course, had never happened, but every member of the crew would have the opportunity to discuss leaving the ship one last time before setting course for Japan. In the harbour, the night shift decided to stay with us in order to see the ship come in and to watch the loading process. They would be able to rest in a spare common room on the *Heron* that was usually earmarked for additional security forces or trainee sailors. The galley would be made available to all security personnel and would be running a buffet for the duration of the loading, which was scheduled to take around five hours – one hour per flask.

It was nice to see the *Heron* coming into Cherbourg. It had only been two days since I had watched her depart, but this major milestone was a great relief. Everything was as it should be and I felt a sense of satisfaction that the planning team, Matt and Bones, had done their jobs so well; and that I was able to contribute to this component of the operation. I could see the balaclava-clad officers on the deck and bridge wings, keeping a lookout as the ship made its way in. The bunting was out on the masts, communicating in an unknown language the nature of the vessel and its intentions. I could see the flags blowing in the wind and the various radar and sensor arrays moving, knowing full well that Chief Inspector Ed, Inspector Mills and several of the officers would be monitoring the whole operation from within the ship's security station. Our shoreside VHF was tuned to the same frequency as the ship, so once we were within sight of each other it was possible to start communicating with them as well.

'Welcome to Cherbourg,' the superintendent announced to them. He had arrived the evening before and was now in the French command vehicle along with an English-speaking French police officer, a luxury we did not have the day before.

Heading to the command vehicle, I saw that the superintendent was getting on well with the senior officers from the gendarmerie. They were sat around the conference table with coffee and snacks while the rest of the personnel saw to their tasks. Rank and status obviously matter quite a lot in the world of nuclear security.

'Come and have a coffee, Princess,' he announced as I stepped through the door. 'The ship will be another half an hour before your lot need to start work.'

I took a disposable cup from a hand that offered it to me. The coffee tasted very good. Much better than I had been expecting.

'He's called Princess,' said the boss, gesturing towards me.

All I was intending to do was let him know that the night shift had decided to hang around – something that didn't need a radio call. The radio channel was being kept clear in case it was needed to coordinate the escort team.

'You are a prince?' came the bewildered response, along with a look of unease from the English-speaking French police officer who was starting to stand up out of his chair.

'No, he's Princess,' came the reply, but not from me.

Confused, the French talked among themselves for a while and changed the subject soon after.

'The Japanese delegates will be arriving later. They own the MOX and are basically paying for the whole thing. However–' he pointed a finger '–they can only go aboard once the loading is complete. We don't want them getting squashed or injured or anything like that. Only the people on the list. Then them, if they want. Understood?'

I had finished my coffee and threw the cup into the bin. 'Yes, sir.'

The flasks for the MOX (mixed oxide fuel) are huge steel cylinders, sat in large cradles that allow them to be lifted using specially designed cranes. Around the size of a shipping container, they are almost indestructible, having been subjected to the most rigorous of tests.[1] They are also designed to prevent radiation from being emitted, so are necessarily made of very dense metals. As a result, they are also extremely heavy. After the third flask had been loaded, the captain and first officer of the *Heron* called a halt. The ship was getting lower in the water and they wanted to wait an hour for the tide to raise the ship into a better position in the dock. The gangway, which had been at a steep angle when it was lowered, was now completely flat, showing just how much the ship had shifted. If they didn't wait for better tidal conditions,

---

1 In one test they were hit by a speeding train. The locomotive came off worse than the flask, which had a few marks on its white paintwork and a couple of dents.

then nobody would be able to get on or off the vessel using the gangway. They would have to use the ship's boat or a helicopter – something that could easily be avoided by just waiting.

Kev and I stood on the dockside and watched the crane retreat along its rails and into the warehouse. It felt strange being a UK police officer working in a French port. We had a small security booth at the foot of the gangway. Walking across to the dockside, Kev shouted to one of the balaclava-clad officers patrolling the deck.

'Who's that?!'

'Kenny,' came the response.

Being level with the officers on the deck made it easier to talk to them.

'Alright, Kenny, how's it going?'

Most of the team were on duty for the loading process.

'Yeah, it's going okay. Looks like the ship's sinking though. The first officer just piped up and told us there's a delay. I cannae wait to get off watch, I've been on duty since last night.'

If Kenny hadn't been wearing a balaclava the signs of fatigue would be written across his face, but he appeared switched on and alert.

'Got some landsmen[2] in their bunks so a few of us our taking extra tasks.'

Kev laughed. 'Who's sick?!'

Kenny shook his head. 'I'm no gonnae tell, it wouldn't be fair on Andy after he bigged himself up so much!'

Andy was not a new member to the escort group, but this was still his first voyage. We knew about him in the planning team as he would often put in requests that were outside the scope of the

2 'Landsman' is a naval term that often refers to new recruits with little or no experience at sea who may be prone to seasickness.

operation – and then threaten to go to the federation if he didn't get his way. After the voyage was completed, for example, he wanted Escort Ops to pay for his flight to Australia, where his girlfriend was on a working holiday. We could only arrange flights back to the port of origin and could only apply for seafarer visas on behalf of the crew. Anything he wanted to do with his own free time was up to him, but it didn't stop him complaining. He even complained about the mattress in his cabin being too firm, and said a mattress topper should be provided. He had, however, excelled in every part of the pre-embarkation training, and was highly qualified, making him a useful officer to have aboard. Strong and intelligent, he could be assigned almost any task onboard – so long as he wasn't seasick.

Heading into the galley, I saw a generous buffet laid out by the ship's cook. The gangway was closed until the ship was sitting higher against the dockside, so I decided now was a good time to take the weight off my feet and get some refreshments. Some of the escort crew were sitting at the tables, as well as some of the crew. Herve was sitting with some of his colleagues as well, so I decided to join them.

'What do you think of the food?' I asked.

They looked at each other, then at me, then at the food and finally at me again.

The feedback wasn't complimentary, nor was it overly offensive. I didn't hear the word '*merde*' more than twice. To be fair, the grub had probably been sitting out all day, and the air conditioning had no doubt taken a lot of the moisture away.

'British food has a reputation,' I said. I wanted to say something in defence of the ship's cuisine but would have only embarrassed myself with my attempts.

We chatted idly until my radio came to life. They were about to resume loading. I stood up and was about to say goodbye to the gendarmes when Herve pulled off one of his Velcro unit badges. He was about to hand it to me but stopped just short of my waiting palm. He pointed at the CNC badge that I had on my shoulder. I thought it was a fair trade, and the transaction was completed. I left the galley and made my way back to Kev on the dockside. With a little luck, the final flasks would soon be aboard and we would be watching the ship depart.

Picking up a copy of the newspaper before heading back to the UK, I read a short article to the rest of the team while relaxing in a lounge on the ferry.

'Meanwhile at the port of Cherbourg, operations took place without the slightest hitch. Just before 11 a.m., the five fuel packages were loaded into the holds of the *Heron*. Various radiological surveys were then carried out by Areva, by the British crew and by the Japanese owners of the cargo. The ship left Cherbourg at 6.35 p.m. and was joined offshore by its counterpart the *Pintail*. The two ships sail together with English Commandos onboard equipped with cannons on the deck. Operation Accord as reported by the French newspaper *Le Monde*.'

# 20

Everything had gone well with Operation Accord. Returning to HQ, I decided to visit the control room to check on the progress of the ships. Behind a solid armoured door, and with strict access control, the CCC (Constabulary Communication Centre) is the national control room for the CNC. Unlike the BCU control rooms that monitored the various sites and units, the CCC had a lot more work to do and was staffed by specially trained officers. One of those tasks was to try to regularly establish and maintain communication with any ongoing deployments or escort ops. This could be done by radio, satellite, through encrypted communication devices or, as a last resort, by fax. There was always an inspector on duty, as well as several PCs and a sergeant.

Gaz, one of the PCs, saw me waiting to get into the CCC through the security cameras and buzzed me in.

'Aye up. Come to check on the ships?'

Gaz was from Manchester and a keen football referee in his spare time. He didn't look like any of the referees that I had seen at football games, being quite round and short. But he had also officiated some high-profile games and was on first-name terms with more than a couple famous football players.

'Just wanted to see where they are against where they were predicted to be by now, Gaz. Good weekend?'

It had been a good weekend apparently. Something to do with the Referees' Association. I felt pleased for him, even if I didn't really understand what he was getting at.

While he was chatting away, I took a seat at one of the stations and logged into the secure system used by Escort Ops. The data for the system was recorded every night onto computer drives that were then retained for a certain amount of time. The system contained not only basic information about operations but also sensitive data that had been passed to the CNC by partner agencies around the world.

The shipowners had plotted a rough route for the vessels to travel, but ultimately the captain of the *Heron* set the course.

'Looks like they're in the Bay of Biscay.'

'Where's that then?' replied Gaz.

'West coast of France. Places like Bordeaux and La Rochelle, you know?'

From there, it looked like the captain was going to take the ships into the deep ocean. *They won't be seeing any land for a few months now*, I thought to myself. The next time the ships would see land would be as they arrived in Japan.

'Matt's away training,' said Bones. With Ed on the *Heron* and Matt off at training, he was now the highest-ranking officer in Escort Ops and had to run the department. I didn't envy his position. He had a lot to do and was not really being very well compensated for it.

We still had to get the details in place for Operation Asquith.

'Why is it called Asquith?' I asked.

'It's a British car, like we always use for job names, and it begins with A.'

'Wouldn't Aston Martin be cooler though?'

'It might be, but I'm calling it Asquith.'

I sat at my desk and looked at my list of things to do for the op. I was pleased to see that my role had changed from earlier; or, more precisely, it had evolved. I was still going to be the 'control tower officer' – a role that didn't seem entirely necessary to my mind – but I was also going to help with some of the reconnaissance work that was due to take place several days before the arrival of the US Air Force. I saw that Bones had given himself a job as well. I thought he would stay in the office and run comms like he had done for Accord, but this time he had appointed himself 'CTSU liaison'.

'What's CTSU liaison?' I asked him from the other side of my desk.

He was reclining in his chair and rubbing his eyes, no doubt from the strain of staring at his computer screens nonstop. Still facing the ceiling as he replied, he said, 'It's a reason to get out of the office. I've been sitting behind one screen or another for almost a year so I'm getting out on the ground this time.'

That was fair enough. Matt, Ed and I had all been deployed in some form or another by now.

The CTSU, or Counter Terrorism Search Unit, was made up of volunteer officers from around the CNC who were trained to search, in fine detail, any building, location or vehicle for bombs, surveillance devices and anything else that might indicate an unhealthy interest in the goings-on of the constabulary. They had officers from the dog section to help them, of course, but the CTSU officers also had tools that allowed them to examine inaccessible and difficult-to-find places. They were due to arrive a few days before the Americans in order to search the buildings that would be used.

'You not running the comms then?' I enquired, scanning the list for familiar names.

'No, Dounreay will handle that. And they're giving us three ARVs along with some extra bods to secure the perimeter. We just need to focus on the inside, and Dounreay will handle the rest.'

Dounreay's CNC contingent was trained in Scots law as opposed to the laws of England and Wales, and while there was a lot of overlap and similarity between these two legal systems, it made a difference. Considering the job took place up there, it was a good idea to have Dounreay officers providing the perimeter security. When it came to the movement of the material itself, however, the Energy Act applied – which is UK-wide legislation.

'The biggest obstacle right now,' Bones added, 'is that we need to get the escort vehicles from Sellafield to Dounreay. And the kit. A lot of kit.'

'Okay, what do you need me to do?' I knew that a task would be coming my way. Whereas Bones had been stuck behind screens for almost a year, I had barely been in the office. It was starting to get tiring, but I reasoned that; a) this is what I had signed up for; and b) maybe this was just an unusually busy period for the constabulary.

'Take one of the recce vehicles and head up to Sellafield. Jump in on the training where you can.'

That didn't sound too bad. But then he added, 'After that you and some of the Scottish drivers can convoy them to Dounreay. That's going to be a long trip, I'm afraid.'

That trip was around 400 miles. Scotland is a deceptively long country. If that was to be my next assignment I would need some time to prepare. I also wanted some time at home. I didn't want to bother Bones with a question-and-answer session, so I sat down at my desk and started planning a schedule. I decided that the best thing to do would be to call Matt, who was commanding the operation, and find out what the REG were doing, and where and when. Then I could find a suitable place to slot in the journey.

I found myself back at Sellafield a few days later and was surprised to see Jimmy Guns there as well.

'You're not back on shift then?' I asked. Unbeknownst to me, he had indicated to Chris and some of the higher-ups during Operation Accord that he would like to move across to firearms training. Seeing as he was technically still on the operation along with the rest of the reserve crew, instead of going back to shift he had been allowed to shadow the firearms instructors and act as an assistant.

'No mate,' he said. 'Mrs and I found out a couple of months ago that she's pregnant. Ten weeks along now. If I had brought it up maybe I would have been taken off the job. Now it's over I want to get into regular hours.'

I congratulated him. I could see a warm glow in his expression. He often talked about his wife, and a move to firearms training would provide him with regular hours and allow him a more 'normal' existence with her.

'Why are you back up here? You spend more time up in Cumbria than down there. Makes me wonder why they moved Ops down to Culham. You'll get a Cumbrian accent if you aren't careful.'

I was thinking the same thing – not about the accent, but the practicality of having a planning department so far away from the centre of most operations. A lot of activities could be achieved remotely, I guessed, but Escort Ops was not a remote job for the most part.

'I'm jumping in with REG,' I told him.

'Ah, well, you can call me "staff" then,' he said while nudging me with his oversized shoulder.

Everybody in the REG was an advanced driver, meaning that they had undertaken a residential course in pursuit and advanced driving techniques. I was only response trained, which meant that I would not be able to drive the armoured BMWs that the CNC kept for this sort of operation, although any response driver could operate them in a pinch. From the outside they looked like any other police car, but the vehicles were heavily armoured and had bulletproof glass in all the doors. They had run-flat tyres and powerful engines that were designed to cope with the extra demands placed on them. Our reconnaissance cars, however, were regular vehicles, with retrofitted police radios and lights. That was what I would be operating before heading to the control tower at Wick Airport.

The large roller doors to the skills area were open, with the police cars and my reconnaissance vehicle parked outside. Inside, some of the firearms training team were setting up barricades and laying out weapons for a training scenario. The 'convoy' was going to practise an ambush from the left-hand side. I stood inside on the raised walkway and looked out at the officers gathered around the vehicles. I could make out Matt, who was the acting inspector for the operation, briefing his sergeants. We looked down and saw a quite complex ambush being set up. The firearms instructors, who were acting as the terrorists, were wearing military-style clothing and carrying AK47 assault rifles and dummy grenades. They had an ambush group and two cut-off groups. It looked very professional and reminded me of an ambush in Iraq that I was lucky to have survived.

I hoped that the REG would have a smooth and easy run – after all, I was a part of it. The portrayal of armed police in the media often focuses on the thrill and danger of police work. However, when real-life situations escalate to gun battles, it often indicates a breakdown in the primary aim of the operation: prevent conflict and uphold peace. Having been a part of several battles, I wasn't keen to relive the past but in a police uniform.

It was September 2004, and we were coming to the end of our deployment in Iraq. We only had six weeks to go. We had recently lost one of our number to hostile action while he was acting as top cover,[1] and another had returned to the UK owing to health concerns that could not be treated in theatre. It was a cooler night than usual at under 40°C, and our two-vehicle callsign was conducting a night patrol through one of the suburbs of Basra known as the Alquasm Market. We were operating Snatch Land Rovers, which offered very little in the way of protection and were sometimes referred to as armoured coffins. Still, it was more armour than we had while on foot patrol. Another issue was that weapons could not easily be mounted on the Snatch, meaning that we were only armed with our individual weapons – L85A2 assault rifles. A lack of equipment had been an ongoing concern for the army over the entire Iraq conflict, and some of our people were still wearing green fatigues instead of the sand-coloured uniforms that we were supposed to wear.

The patrol commander, a lieutenant, was leading the patrol from the front vehicle and had Drewey as his driver. Drewey and I had been through basic training together and were happy to have been put into the same multiple.[2] My vehicle was commanded by Fat Rob – a nickname bestowed by the local Iraqi kids owing to Rob being as wide as he was tall. We had Stick driving our Land Rover. Stick was not thin and wiry, as his name might have suggested, but instead had the agility of a stick – that is, very little. Along with Broz, a Gloucestershire boy, I was acting as top cover.

As our vehicles crossed into Alquasm Market there was an explosion. The blast wave pushed me to the side, causing me to

1 Top cover – the soldier assigned to mount the weapons or stand as a lookout while operating in a vehicle.

2 A multiple is a subunit of a platoon that is larger than a section.

lurch over the open hatch that I was standing in with Broz. Stick put his foot down and all of a sudden I was thrown the other way as he swung the vehicle to the side in an attempt to escape the ambush.

'CONAAAAACT!!' shouted Broz as he started shooting.

I picked myself up and looked to see where he was firing. Before I could identify any targets I saw another explosion behind us. It seemed slower than the first one and it looked white – not like the orange Hollywood fireballs so often seen in films. It was close, too. The first explosion had been between our two vehicles, which were only about 20 metres or so apart. Behind us a wall was disintegrating from the impact of the RPG[3] that had been fired at us. With two targets to fire at, it was a miracle that neither of the vehicles were hit. A second RPG had been fired at the rear vehicle, which I was manning. The rocket must have hit a tree or lamppost or piece of street furniture as it seemed to explode low and without hitting anything obvious.

We started to take small-arms fire. The AK is an extremely loud gun that fires 7.62mm bullets at a high rate. There were bursts coming from all around us, indicating that the ambush had at least a few groups. Forcing myself to focus and return fire, I tried to look for the source of the bullets coming my way. I wasn't scared at that moment; instead, I was suddenly very frustrated. I felt like I was the only member of the patrol that wasn't doing anything. Maybe I was. The drivers were driving, the top cover were shooting, and I hoped that the commanders were commanding.

I saw a figure running on the other side of a canal that separated Alquasm from downtown Basra. Then another joined him, and a third. They were repositioning. Clearly they were armed. I could

3 Rocket-propelled grenade.

see the sickle-shaped magazines protruding from their weapons. One was carrying a tube on his shoulder. This must have been an RPG or something similar. It looked like they had broken cover and were trying to set up a shot to the rear of our Land Rovers as we made our way out of the contact on Tamooz Street, which had a long, straight section. We would be sitting ducks.

Broz was still firing. I could feel his ejecting bullet cases bouncing off me while I took aim. My sight picture was shaking from the evasive driving but I had someone in my sights. It was one of the three. I fired. It wasn't a single well-aimed shot, but five or six rapid pulls of the trigger. Bangbangbangbangbang. He went to ground. Did I hit him? Did Broz hit him? Was he taking cover?

Before I could fire some more rounds, we turned hard left onto Al Saadi street. I struggled to regain my balance and immediately began looking out for any follow-up ambushes. I could hear on my radio that we were making our way rapidly back to Basra Palace. There was now only the sound of the engines – as well as a ringing in my ears. I turned to face forward. We were close behind the lead vehicle. Both of their top cover were still standing, and looking out. I continued to look around, now hypervigilant, but then remembered the sensation of the bullet cases from Broz's gun hitting my chest. Were those bullet cases I had felt? I quickly looked down to make sure that I hadn't been hit.

I hadn't been hit. Nobody had been hit. It had been close.

It looked as though the firearms instructors were setting up a similar ambush for the escort group in Sellafield. Except this time, the REG would disembark and take on the threat. There was no way a slow-moving nuclear transporter could escape quickly. The heavy vehicle was built with security in mind but also utility. It was designed to haul dense metal containers that contained the most toxic material on the planet. The best defence for the officers escorting the cargo was to get close and engage. This was one of the defining differences between the MEG and the REG. The scenario-based training that the REG undertook reflected that, as did their equipment.

I leaned on the railing of the raised walkway and watched Matt's team go to work.

# 21

The team finished their exercise. It took no more than a minute, and it looked slick and professional. Indeed, it looked *too* good; I wondered whether it would appear quite so efficient outside of a training environment. But like with the MEG training I had done earlier, everything was moving from simple to complex. While we had been on our pre-embarkation training, the REG had been revisiting their driving skills. They had to return to shift for a few weeks after that so that an initial firearms course could run through B357. Now they were beginning their training in earnest and had access to the entire firearms training department. They needed it. With a full complement of officers, armoured vehicles and unmarked police cars, the REG took up the entire facility. They even had one of the old UKAEA Constabulary Transit vans to act as a fuel transporter.

I sought out Matt after he finished the 'hot' debriefing and asked him what he wanted me to slot in on. He called over Mick, one of the team sergeants.

'Mick, this is the guy I was telling you about. He'll be in your wagon for the recce.'

Mick stuck out his gloved hand to greet me. 'How's it goin'?' he said in a Northern Irish accent. He looked tough. A short guy in his forties, he had a cheerful look that belied his serious nature. No doubt he was a rugby player, or had been.

'So what we're doing now is refresher training on basic skills and a little bit of work for the convoy itself,' Matt went on. 'I want everyone to know everyone else's jobs. It's repetition, repetition, repetition at this stage.' He pointed a finger. 'After the weekend the recce teams and support teams will take their motors to the diver training venue, and practise there. The close protection will stay here and practise exactly that: close protection. You got your kit sorted?'

A lot of the tactics employed by the REG mirror what might be employed by modern military forces. After a few days of settling in and going over some of the basics, we headed to a nearby MOD site to practise our vehicle tactics on a disused road network within the perimeter. It was the closest the escort group could get to realistic on-the-road training. The reconnaissance and response parts of the team were scheduled to train first, followed by the close escort team a few days later. Using powerful unmarked police cars, the advanced drivers practised their skills, driving at breakneck speeds before rapidly positioning the cars for immediate action. As the CNC has no driving instructors, sitting in the passenger seats of the vehicles were officers from Cumbria's RPU, or roads policing unit. It had been a while since the drivers had been able to practise their craft, and it took most of the morning.

'You fancy a crack at it?' Mick asked me. Like the rest of the officers in this part of the team, he was a skilled driver. I, however, was not. I was only response trained, a tier lower.

Still, I agreed and got into the car with the RPU instructor. When I had been on my response course in North Wales, I knew that there was a 'cockpit drill' to carry out whenever you took over a car, but for the life of me I couldn't remember what it was. Instead I just did what I always did, and adjusted all the mirrors and the seat – hoping that it would be enough. I was very used to the vehicle anyway. I had been driving it a lot, going from venue to venue as an operations planner.

The instructor set out what he wanted me to do, which was the same thing the others had been doing: navigating a simple obstacle course and then flooring it to get to the action point. This would be the point where the officers, had we been carrying any, would disembark and go about their business. Simple enough. So that's what I did.

'Faster, come on, faster.' The instructor was urging me on. Clearing the network of cones that had been set out in sequence, I put my foot down to head to the other end of the road. I had barely worked my way up to third gear before starting to slow down again, bringing the car to a sudden halt that pushed me into my seat belt.

'A bit timid, aren't you?' the instructor remarked. 'Fancy another go?'

I looked out of the window and could see the rest of the officers in a group, talking with Mick and some of the others looking this way.

'Okay.' A one-word answer was all I could manage.

'When did you do your advanced?' the instructor asked as we drove back to the start point.

'I haven't. I'm response.'

'And being CNC I bet you've never even driven a response?' he queried, already knowing the answer. He was right of course. 'I didn't know response was allowed on escorts. Oh well, never mind. Drive back to the others.'

When we returned and I got out, Mick came up to me. 'Need another go at it, do you? That was slow. And we all saw it.' No compliment sandwich from Mick, then. He was right.

I was certain that the other officers needed little instruction or encouragement.

Each of the several vehicles we were using had a team assigned to it. The number of officers changed depending on the kind of car that was being used and the area in which it was being used. For this reason, the road escort officers needed to be quite flexible and able to work in a number of roles. Their initial training reflected that. I had not done the initial training and, being assigned to route reconnaissance for the operation, it was not entirely necessary that I did it. The team, however, needed to be able to deal with any threats that they identified while carrying out their patrols. This meant that the officers were fully equipped in much the same way that they were during the actual escorting phase of the operation.

We mounted up into the car. Mick was in the front passenger seat, alongside the driver. I sat behind the sergeant along with another officer in the back seat. That meant if we dismounted to take on a threat, Mick and I would form a pair while the driver and his opposite number would form a second pair. No doubt Mick wanted to keep an eye on me should anything happen. Either that or he didn't trust that I could do the work.

'Contact drills, then,' he announced as the doors shut.

I looked around to see if anyone was wearing a seatbelt. They weren't, so I left mine off as well. The firearms instructors had set up some Elvis targets at some point along the track and were going to initiate the 'contact' with a burst of blank fire from an AK47. It wasn't hard to guess where the targets had been placed, as the officers still waiting their turn were milling around a little way up the road. What we didn't know, however, was the number of targets or their distance from the road.

We headed down the road. We were on some MOD property that had been put into a 'care and maintenance' state, meaning that it could come back into service at some point. The REG, however, had been using it on and off for years along with certain parts of the military such as the Royal Military Police and special forces. The camp part of the site looked like so many others I had visited, with lots of single-storey buildings and an administration block. The road network that worked its way out of the camp went into some open country that was presumably at one time used for storage of heavy machinery or munitions. This meant the roads were identical to those in the rest of the UK as they were designed to withstand constant heavy traffic.

The vehicle built up speed and soon we were cruising along at a sensible 30 miles per hour. We could see the other officers up ahead.

'Switch on then boys,' Mick said as we started to draw level with them. Then past. No contact, no nothing.

'Keep a lookout,' the sergeant said. He was looking in the mirror as we went past the observing officers.

Pretty soon we had completed another lap. Then another. As we joined the road to start the fourth lap, we got the signal. We could see the instructors on the right side of the car – the driver's side.

'CONTACT RIGHT' came the calls, in unison, from all of us, and the car quickly came to a halt. I disembarked from the left with Mick, who leaned on the bonnet of the car and began shooting. I peeled around him and got down into the prone position.

'COVER ON!' I called as I started to engage the target that I could see. The blank rounds we were using weren't overly loud, but it was still necessary to shout to make communication possible.

By now the driver and his oppo were in position and had begun firing as well. Mick bound past me to a wall made from old tractor tyres and began to open fire again. 'COVER ON!' he called.

I made my way past him, and we repeated the process until we were called to stop by a firearms instructor.

'Endex!' said the blue god. I had no idea who he was, but he seemed to know the other officers. 'Noice,' he said to Mick in a very casual way. No feedback though. Just a few questions to check that we knew what we were doing. By now we were a good 50 metres away from where we had started. How many targets were there? How many rounds did you fire? What time did the contact start? It was policing 101.

# 22

We continued our build-up training back at Sellafield. Matt and the rest of the team were using the MOD site, meaning that there was plenty of space in which to train. The instructors, with Jimmy Guns in tow, were running us through scenarios that we might expect at Wick Airport, the trans-shipment point.

Based on the design basis threat (DBT), the training focussed on how to respond to a well-armed attack on the airport. The International Atomic Agency Authority had conceived the DBT in response to international security concerns brought about during the War on Terror. Essentially a detailed threat assessment, the idea is to have the security resources in place to deal with 'external adversaries' who pose a risk to the nuclear industry, including terrorists. Once done, in a few days, the logistics part of the operation would need to begin. I was not looking forward to that. I had been busy training alongside the REG and enjoyed being out of the office, but technically that was not my role. My upcoming task would be to get a convoy of heavily armoured police cars from Sellafield to the top of Scotland.

By now Bones had joined me and Matt, meaning that we could begin to consider the route the escort was going to follow. As the operational commander, Matt would have the final say before passing the plan to Wally, who was in overall charge of anything escort related. I could see Bones on the gantry above the skills area, taking some pictures of the training with a digital camera. I nudged the officer next to me and pointed up towards the viewing gantry.

'Smile for the camera,' I said while we both waved and struck heroic poses.

I knew that I would have to excuse myself from the balance of the training in order to resume the planning of the operation, but I completed the rest of the scenario.

'I've got to go to see what the planners are up to, sarge,' I told Mick before the next serial of training began.

'Oh yeah? Had enough have you?'

'I think it's time to plot the route and arrange the logistics. Get everything moving and get everything ready,' I told him.

He nodded. I think that was the only approval I was ever going to get. 'Let me know when you sort the route out,' he added as I began to walk up the gantry.

At the top of the gantry I greeted Bones's nod and unclipped my helmet. I had been sweating, and the breeze travelling through the skills area made the top of my head feel uncomfortable.

'I've moved us into some temporary offices in Summergrove,' Bones began. 'We've got everything we need for the final bits and pieces. How are you getting on with the PAR?'

In between training I had been taking equipment requests and packing orders with Marie, leaving very little of my PAR actions outstanding.

'I've got a few more things to do. It'll be fine.'

On a clipboard he had a printed copy of my PAR actions, and going down the list we marked them off as either 'completed', 'in progress' or 'not started'.

'Have you spoken to the CTSU, found out what they need?'

I hadn't. I had forgotten about them entirely and I didn't even know who their skipper was.

'No, sorry.'

'So whilst you've been training and having a good time, basically you've sorted your new friends out with kit and not even started to see if the CTSU has everything they need?' I assumed that was a rhetorical question. 'I suggest that you go and start those actions now.'

He handed me a printed list. I could see the stress on his face. He had been left with a lot to do; more than was reasonable, especially with the other two planners, Matt and myself, out of the office.

'I'll do that now,' I said, more to the list in my hand than to Bones. 'We're in the training centre's B wing offices. I'll see you in the Moon Bar later.'

I dekitted and packed my bags, chucking them into a spare car that belonged to me for the week. Then I drove to B520, the CNC Sellafield police station. I knew that if I wanted to get hold of the CTSU then I would need to go there and find the control room directory.

The control room buzzed me in. It was smaller than the CCC at Culham but was designed to respond to the needs of the site rather than the entire constabulary. It was busy, though. There were more people working there than at the CCC.

'What do you want?' asked an inspector, not impatiently but not politely either. I was still dressed in my coveralls and boots, with sweat marks around where my equipment had been.

'I need to get in touch with the CTSU sergeant,' I told her.

'Why do you need to get in touch with him?' she asked.

'Nothing major, I just need to find out if his team has any equipment requests for the road escort job they're attached to.'

'Someone get the CTSU sergeant for him, please,' she said to nobody in particular.

Finally, I sat down and was handed a phone handset.

'Dounreay firearms training,' said the receiver in a soft Scottish accent.

'Hi mate, Escort Ops here. Have you got the CTSU sergeant over there?'

'Aye, wait a second, I'll get him.' I heard the phone receiver tap on the table as he set it down, and then, 'Alistair? Some guy from Escort Ops on the line.' Then the fumbling of a phone being picked up.

'Hello?'

'Hi, mate I'm trying to get hold of the CTSU skipper.'

'Oh? Well you've got a hold of him. Can I help you with something?'

In an age where email and mobile phones were so common, I didn't expect to go to all this effort to do an equipment check.

'Yeah, it's PC Okuhara from Escort Ops. I'm in the process of doing equipment issue, so need to check whether you have everything.'

'No, we're fine.'

That was a short answer and didn't really satisfy the task I had been given. 'Fine' by his standards and what the operation required might be two different things. I took the PAR sheet out of my leg pocket and started going down the list, asking about specific items and logistics issues.

Writing notes onto a pad that I always carried with me, I found that 'fine' was actually almost good enough. They didn't have a van to move everything they intended to bring, but all the technical equipment they needed was in a storeroom in Dounreay. That was lucky. Maybe Bones had moved everything up there and not ticked it off the PAR.

'Oh, and can we get bacon rolls for breakfast? It's an early start, you know, and we'll be working all day.'

Bacon rolls? Was he serious? I was calling to find out about technical equipment and any PPE they might still need, not refreshments. But with my dressing-down from Bones ringing in my ears, and the indifferent inspector sitting only a few feet away, I decided against going into it.

'I'll see what I can do.'

Back at Summergrove later that day, I joined Bones in the offices that he had set up. Like when I had been there as a recruit, the offices were converted bedrooms. He had brought several laptops, a portable printer and a device that allowed the small offices he set up to connect securely to the CNC network.

'All sorted,' I announced as I walked in. 'I need to speak to Amy or Candice to order a few cars and a truck, but other than that the list is looking good.' I handed him the sheet he had given me earlier, with ticks and annotations as evidence of my hard work that afternoon. 'I was surprised that the CTSU kit is already in Dounreay,' I added.

'Why?' he asked.

'Well, I just thought that it was lucky that we don't have to move it.'

'The CTSU team stores have always been in Dounreay. There's no luck about it. The CTSU training course is in Dounreay and the CTSU sergeant is a Dounreay firearms instructor.'

That explained that then. I sat down and got on the phone without replying to a clearly overworked skipper. After all, three of the department – Ed, Matt and myself – were doubling up as operational officers. There had been a round of temporary promotions, and we had been going nonstop since the office had moved down to Oxfordshire. Now here we were, back in Summergrove, where ops planning had previously been located, doing the role from here instead, albeit temporarily.

'I've had enough for today,' announced Bones. 'I'm going for a walk. You coming?'

It didn't sound like a bad idea.

We sat in the Moon Bar that evening. The planning team had been so spread out and busy that most of the planning had been done by the temporary sergeant.

'I'll get them,' volunteered Matt. Setting down three pints, he inhaled deeply and looked at us. 'All okay?'

Bones answered immediately. 'You're the inspector for planning, and he's only a PC, but you're both up here playing at road escorts, when you should be planning with me. It's only just okay.'

As always, Matt showed very little emotion whereas I took an extended swig from my drink.

'Well, I do have some good news for you, but act surprised when Mr Watson speaks to you after the job. You're going to be promoted substantive sergeant. If you want it.'

That changed his mood. He was happier now.

'What about you? That's your role.'

'I'm out of here after this. Keep it to yourselves as well, but I'm going to be the acting inspector for special branch when Ed gets back.'

That was good news for him too. I hoped there was some good news for me, but there wasn't. Nothing was brought up and nothing further was said. I would be back to Escort Ops.

'Who's the new planner going to be?' I asked.

'Fern. She's welfare officer at the moment.' I remembered her. PC 138, who had commandeered my desk while I was away. It was probably a good thing to have her in the office, as everyone spoke very highly of her. Not only that, but she was also the only person in the constabulary who could stand up to the superintendent's explosive temper.

'Back to business,' Matt said after a drink. 'Tomorrow and the next few days will be route planning, printing the operational orders, and then you need to drive up to Scotland. After that, the planning is over and the op begins.'

The cargo that we were to escort needed to travel from Dounreay to Wick Airport. That was a distance of around 30 miles.

The superintendent and firearms instructors had suggested some routes and provided Matt with a lot of information to help him decide which course to take. He was keen to avoid settlements like Thurso and Castletown, but they had roads more suited to a convoy of heavy vehicles. A rural route would involve more twists and turns, and potentially be slower as the transporter would constantly need to bleed off speed on the many bends in the road. Matt wanted to avoid Loch Watten as well, and steer clear of any major bodies of water, as if the transporter crashed or was hijacked then it would be all but impossible to recover it from a toxic lake.

Eventually he came to a decision, and in the office we set about finishing the operational order. It was smaller than the orders that had been put together for Operation Accord, but it was still highly confidential, so every copy had to be signed for and accounted for at all times. Once again, that meant me sitting by a printer, taking the copies of the operational order and keeping them safe until they could be distributed.

Finishing up, I made myself ready to drive up to Scotland. I would be driving one of the vehicles, and the others in turn were to be driven by officers from Dounreay who had been taking part in the road escort training. The rest of the team would fly up a few days later.

# 23

The response officers had their briefing first. Similar to the military, they were to act as a QRF, or quick reaction force. They would respond to anything that was sent their way, from towing off broken-down vehicles to dealing with protesters on the route or providing an armed response to threats. To accomplish this, like those of on reconnaissance, they used unarmoured but very powerful unmarked police cars. They were given predetermined points at which to wait, and these changed with the phases and timings of Operation Asquith. To begin with they were to patrol around the perimeter of Dounreay site, adding additional external security to the start point of the operation.

My team, meanwhile, was to commence the route reconnaissance a considerable time before the convoy was due to depart. This required driving the route and checking every potential obstacle that had been highlighted in the briefing: things like bridges, corners where the convoy would need to slow down, and crossroads where there was the potential for traffic to split the convoy. We were also to look for any suspicious activity that might suggest an interest in the convoy route, as well as any unattended cars that didn't fit the profile of the area. It was going to be busy and laborious – but it was a necessary part of the escort.

After completing the first recce, I left the team and took one of the spare vehicles. Out of curiosity, I decided to drive the route Matt had selected. My place in the team had now been handed to a Special Branch detective who was also part of the operation. Driving the route at the speed I thought the convoy would use tomorrow, I found that it took a lot longer than I had anticipated.

I parked outside one of the emergency gates on the perimeter of the airport track and made myself known to the two CNC bobbies manning it. I smiled, remembering the nonstop jokes about the CNC and gates. No doubt the best gate guardians in the world. They let me in, and I drove around the taxiway to the terminal building. Nearby, there was an innocuous store building that we were going to use to unload the cargo before loading it onto the US Air Force C130 Hercules cargo plane. Most air traffic had been halted, particularly recreational and training flights, although the odd commercial aircraft was still using the airport as scheduled.

Inside the store building the CTSU were at work, using their special equipment to search parts of the structure that had probably not been looked at since it was erected. There were air conditioner parts on the floor where some of the officers had taken the casings off the machinery to get a better look inside. It looked like they knew what they were doing. I hoped they did; I wouldn't be happy if someone took my property apart and then put it back together incorrectly, or broke it. I couldn't remember seeing an insurance policy anywhere if they messed it up, either.

Bones nodded and called me over. 'The CTSU skipper said something about bacon rolls,' he said in a quiet voice so as not to interrupt the search.

I had forgotten about that. But I also remembered that I hadn't agreed to anything either. Thinking that would be the last I would hear of it, I said, 'Nothing to do with me sarge. I'm a vegetarian as you know.'

We went outside. 'I'm not impressed with how the CTSU search is going,' Bones confessed. 'Only half of the team seems to be working at any one time, whilst the other half is in Tesco drinking tea and eating cakes. I reckon we would have finished by now if the whole team was in there.' I knew by now that Bones was a very practical and straight-talking guy, so what he said was more than likely something to which he had given a lot of thought. 'Really we should have finished by now.'

There was no 'set' time to complete the search of the store building, so long as it was completed before the cargo arrived. However, there were officers from Dounreay waiting to secure the building once the search was completed, and they were yet to be deployed.

'It is a big building, though,' I offered in their defence.

'Yeah, but there's fuck all in here. How were your recces?' he asked, changing the subject.

'I only did one. Someone from SB is riding along for the next. But it all seems good. I'll leave you to it and introduce myself to the control tower people.'

I had not been in an air control tower before, but it looked how I had imagined. I was given a magnetic keycard that let me in and out of the door at the bottom of the tower, although it looked like a good tug would probably break the hinges off. There was only one person working. She explained to me that she and the other controller took it in turns doing two-hour shifts, and that she was the one expecting to be on duty when the Hercules transport aircraft arrived.

I introduced myself, but not by nickname. She was a pleasant lady and very chatty. This was her first career after her kids had flown the nest, she told me. She had taken advantage of a Highlands and Islands Airport initiative to get more women into aviation operations, and by all accounts she was loving it. It sounded great. I knew very little about the industry myself, but she was certainly doing a good job selling it to me. Pleasantries over with, I told her about what she could expect tomorrow, and what my job would be. It didn't seem to faze her in the slightest, as if it was only a little training aircraft practising touch-and-go landings and not a sensitive multinational operation.

'Okay then, I'll see you tomorrow,' she said as I made my way back down the stairs. I thought to myself that it was probably a 'do nothing' job. Maybe even a role for the sake of a role. But being high up in the control tower gave me a premium view of the goings-on.

The next day, with the strategic and tactical briefings over, the escort team assembled their vehicles around the transporter. A custom-built armoured flatbed, this lorry was designed to lift and carry the heavy fuel flasks that are often found in the nuclear industry. The armoured cab serves two purposes: it protects the crew from any hostile action, but it also shields them somewhat from the dangerous cargo in the event of trouble. I watched them arm and equip themselves before mounting up and heading to the rendezvous on Dounreay site. Bones was in the Dounreay control room with the boss, Wally, keeping an eye on proceedings, and once again I made my way to Wick Airport, this time in a police car and not one of the spare admin cars we had.

The officers at the gate waved me in and I parked outside the control tower. Then I joined my new friend in the tower and sat down beside her, ready for the escort to begin. Northern Constabulary[1] had deployed a team of motorcycle outriders to stop traffic at crossroads, and the reconnaissance and response

1 Prior to the amalgamation of all Scottish constabularies into Police Scotland, Scotland was made up of several county-sized forces: Central Scotland Police, Dumfries and Galloway Constabulary, Fife Constabulary, Grampian Police, Lothian and Borders Police, Northern Constabulary, Strathclyde Police and Tayside Police.

teams were already out and about. Once Mick reported in, the convoy would begin to move. My radio was tuned to the escort frequency and I heard Mick contact the control room.

'Romeo Echo Nine, all clear.'

In turn, the control room contacted the convoy and it began to roll.

Loitering off the east coast of Scotland, the USAF Hercules had flown from Ramstein Air Base and tried to time its arrival to match that of the convoy. They wouldn't be waiting long; while they were in contact with the control tower, they were also getting updates from the security people that the American Embassy had sent. I didn't know where they were, or how they were so well informed; I assumed that they must have come to some sort of arrangement with our own people.

Outside the perimeter to the airport, a small group of aviation enthusiasts had noticed, using an online service, that an American aircraft was circling nearby and had come armed with radio scanners and cameras, in order to get a glimpse of the plane as it arrived and departed. I could see them through the binoculars that were kept in the control tower and called the control room to check if one of the perimeter teams should speak to them. Maybe I wasn't entirely redundant after all. I didn't want to clutter the airwaves with chatter, preferring to leave the radio network clear for the officers on the escort, so I called Bones.

'Yeah, alright Princess, send one of the teams to have a look.'

I had been through my basic training with Alan, who was now a sergeant at Dounreay and acting as the skipper for the perimeter teams. 'Delta Romeo One,' I hailed.

'Go ahead.' It was strange to not hear a Scottish accent on the radio. Alan, as he had told us all on Course Five, had left England behind to start a new life in Scotland. 'Delta Romeo One, there is a group of five or six people at…' I gave him the location. 'Can you send someone to check on what they are doing?'

There was a pause. 'Received.'

The convoy would also have heard that transmission, so to avoid causing any alarm or concern I added, 'Delta Romeo One, it looks like some aviation enthusiasts but control wants you to speak to them and ask them to stay well back.'

Another pause. 'Received.'

It wasn't long before I saw a CNC patrol car driving around the perimeter track to the location. The blue lights were flashing, which I thought was a bit much; but then again, he was the sarge, and he probably knew better than me. Through the binoculars I could see Alan emerge, along with another officer, and begin speaking. It all seemed innocent enough, and I was not expecting to hear otherwise.

The convoy reported its position. Soon after, an American accent came across the control tower radio. The Hercules was preparing to land. In truth, the aircraft probably had more sophisticated radar and control systems than the airport and could have landed without assistance from the friendly air-traffic controller. Their contact was also the signal for the perimeter officers to unlock the gates that were being used to bring in the convoy. By now, Alan had rejoined his crew. I saw a pair of police cars at the gate, while another patrolled the taxiway adjacent to the fence.

'There they are.' I pointed to the approaching convoy.

'There she is,' replied the air-traffic controller, pointing at a radar screen. 'Good timing.'

The C130 came in and roared as the engines engaged reverse thrust to slow her down. She taxied next to the store building that the CTSU had cleared and shut the engines down. The ramp came down and a low-bodied trailer pulled by a small tractor disembarked. Soon after, the convoy arrived and the transporter, along with one of the escort vehicles, headed to the store as well. I knew from experience that moving the heavy cargo even from ground to vehicle could take a long time, so this was not likely to be any different. By now the escort group had deployed to intervention points around the airport, while the unmarked vehicles carried on patrolling around the immediate area.

'Won't be long now,' I said. 'Then you can get back to normal.'

It had been a unique operation for the airport, and a unique operation for the constabulary as well. Whatever was in the fuel transporter, the US Government wanted it at quite short notice, leaving us with little time to plan, prepare for and execute a road escort. But as I heard the engines of the Hercules start to power up, I knew that we had managed to pull it off.

# 24

Following the successful completion of Operation Asquith, most of the team returned to their duty stations, taking numerous bags of equipment, crates of firearms and police equipment with them. A few of us remained, however, to debrief the operation and discuss in detail what went well and what could have gone better. Occupying a conference room at Dounreay, the senior officers from our partner agencies, along with Matt and Wally, spoke to the sergeants about the merits of the operation. I was there too – the only constable – taking notes and recording the meeting on a laptop that travelled with the planning team wherever they went.

As I expected, the reconnaissance team sergeant, Mick, complained about having to take a 'tourist' with him, mentioning that Special Branch should not deploy in the same way that the escort group does. He made no mention of having a planner with him, though. The convoy sergeants all agreed that it went well, but they were not happy about the route selection. Matt explained to them his reasons for the route he had chosen, suggesting that even if it took longer than a direct route it also meant that fewer people would be around.

Sgt Williamson disagreed. 'I think we need to think about limiting the cargo's time on the road,' he said. I wrote it down.

Then the CTSU sergeant spoke. 'We had been told we were getting bacon rolls in the morning by that planner over there,' he said, pointing to me while I caught up on some of the previous points.

'What?' The superintendent was nonplussed.

'I had to go to the shops and buy them myself once they opened. We were expecting some breakfast before we started the search.'

Wally was always very direct in his manner. 'Did you have breakfast?'

'Aye, sir but we were told that we would be getting some bacon rolls as a snack.'

Wally looked at me. He looked at the sergeant.

'Your desire for a bacon sammich is not the focus of this debrief. I expect a team of fully grown men to turn up for duty rested *and* fed. It is not the function of the planning team to act as your gophers. Get it? Now. Tell me about the search.'

Thinking better of recording that exchange in the notes, I merely jotted down that 'refreshment issues were discussed'.

Bones chipped in that in his opinion the search took longer than expected but had still been completed within the scope of the operation. It had, however, delayed the security arrangements slightly, although not majorly. There were entries on both sides of the ledger, good and bad, but with the operation having been brought to a satisfactory conclusion, very little would have been changed if it was to be done again.

That evening, those of us still at the hotel celebrated the conclusion of the operation in the hotel restaurant. The wine and beer were flowing, and the mood among the team was good. Standing to give a speech, the superintendent clutched a large Tesco shopping bag that appeared to be stuffed full of items.

'First off, and once again, well done to all involved.'

People banged their tables in appreciation.

'The chief bobby is pleased with what we achieved at such short notice, and so am I. To that end, I have decided to award prizes to all of you here.' That explained the shopping bag. I could see the boss had already sunk more than one drink, but I guessed that he was used to that kind of activity.

'Matt,' he called and reached into the bag, rummaging around to find something. 'Matt, as the convoy commander, I hereby award you this Automobile Association map of Great Britain, so that next time you can select a better route.' We laughed as he handed it to him. 'I've put a bookmark on the page for Thurso and Dounreay,' he added. Matt pulled out the bookmark, which appeared to be homemade and simply said 'good job', written in marker on a folded piece of paper.

'James? Where's James?' Bones was sat beside me and put down his drink, pointing a finger at himself in surprise. 'I hear it was a stressful time for you.' More laughter. 'People up here know how to deal with that effectively, so take this medicine.' He produced a small bottle of Old Pulteney whisky, a local brand.

'I'll save it for the next job, boss,' he replied, taking it from him and returning to his seat.

'Princess!'

I was next. At least I had not been forgotten, although I wasn't sure what kind of embarrassment I would be put through.

'Next time, so that you don't get bored in the control tower, I got you this.' It was a small, pink doll's house about the size of a hardback book.

'Thanks, boss.' I stood to collect my prize as the laughter and good humour continued.

Prizes were given to everyone present. Chocolate medals, bottles of Irn-Bru[1] and, of course, a bacon roll for the CTSU skipper.

With no further operations planned, the Escort Operations department was able to start catching up and returning to a more sustainable pace of work. Along with checking up on the progress of Operation Accord, we arranged for the regular training of personnel and for the maintenance of our equipment, which was held separately to that of the rest of the CNC. I was also due to attend a requalification shoot, and had driven to the constabulary's new shooting venue at Bisley. I thought Bisley was an odd place to have a firearms training facility given the lack of infrastructure for that purpose. Bisley was more suited to hobby shooters who would live in tents or caravans during a competition. The new chief firearms instructor (CFI), however, had arranged to use the facility to eliminate our reliance on host forces' training venues. Over time, no doubt, the facility could be improved to accommodate a sizeable portion of a permanently armed constabulary.

Sitting outside one of the many pavilions dotted around Bisley's shooting ground, I went to check my phone, having felt it vibrate in my pocket. It was Jimmy Guns. Sellafield was in lockdown and a major incident had been declared. An active shooter[2] had murdered three people in targeted killings and was now shooting people at random in the areas around Sellafield and Whitehaven. Jimmy, who was now a firearms instructor, had been pulled off

1 Irn-Bru is a carbonated soft drink that originates in Scotland.

2 Active shooter – a term used to describe the perpetrator of an ongoing mass shooting.

instruction duties along with three other CNC officers to assist in the manhunt for the suspect, Derrick Bird.

While fulfilling some equipment requests at B357, I went to catch up with Jimmy and get his version of events. During a break from an initial firearms course he was working, he gave me the details.

'This guy starts off his day in the shittiest way possible. He heads over to his twin brother's place in Lamplugh and shoots him eleven times. Eleven. It's like something out of a horror movie. But it doesn't stop there. Then he goes by the family solicitor's house. Bird's got a sawn-off shotgun, and it's just chaos. Shots are fired, the solicitor is down, and it's all escalating.'

I knew this much already as the news had been analysing the mass shooting non-stop. It hadn't been until the solicitor's neighbour called the police that anyone knew a rampage was in progress. At this point, there was still a lot of speculation around the motives for the killings: personal differences, financial issues, jealousy, mental health. All of them were being brought up.

'We knew nothing about it until Cumbria called for some support,' he added. 'The deputy CFI called a halt to all the training and ordered us to deploy in two of the Road Escort ARVs. Whitehaven became the stage for his next round of shootings. He headed to the taxi rank where he worked. He takes out a taxi driver. Then more chaos – more shots, more people caught in the crossfire. Imagine being in that taxi rank, thinking it's just a regular day. Then all of a sudden…' Jimmy mimicked an explosion.

Jimmy had given a statement to West Mercia Police after the incident, not as part of the criminal investigation into the shootings but as an independent constabulary charged with reviewing the post-incident process. This allowed the higher-ups in the police to analyse the response that Cumbria Constabulary and the CNC were able to provide.

'Bird's targeting taxi drivers left and right. Some get wounded, some barely escape. Cumbria try to step in, but they're only county bobbies in stab vests so there isn't a lot they can do.'

Cumbria Constabulary was already busy policing the Appleby Horse Fair, which attracts tens of thousands of visitors every year, and was also marshalling a large funeral after a bus crash that had happened earlier that year. Fewer resources were available to respond, and diverting them would take time that Cumbria Constabulary did not have. Heavily armed, Bird was carrying a double-barrel shotgun and a .22 calibre rifle. Even though that's a small calibre compared to many guns, it is still lethal and more than capable of causing serious injury. Within thirty minutes of his attack starting, three were dead and four were injured.

Cumbria is a very large county, and its network of roads can make travelling from one part to another very slow. Police resources are therefore spread quite thinly at the best of times, so mustering enough officers to conduct a manhunt for an armed and dangerous individual can take a very long time. When the CNC was created, the Act of Parliament that created it also included a proviso that CNC officers could be seconded to other constabularies should the need arise. This has happened on several occasions, and the Cumbria shooting was a case in point for those at Sellafield.

'Cumbria sent us out looking for him. We were to "intercept and confront" him,' Jimmy emphasised. 'They didn't have enough people to cover such a large area. Residents were told to stay indoors, the site was locked down and until it was over, the gates remained locked.'

Bird himself had worked at Sellafield in the 1990s but was sacked when he was found guilty of theft from the site. After his targeted killings in Whitehaven, he made his way to Seascale and Egremont, two towns that back onto the nuclear site. In Egremont, he shot at a dogwalker but did not hit her. He then shot at three more people, killing two and leaving the other with life-changing injuries. By now the search had escalated, with one officer from Cumbria Constabulary commandeering a worker's van in pursuit of Bird, who was driving his taxi.

The officers chasing Bird then lost him. It turned out he was doubling back the way he had come, heading away from Egremont and back towards Seascale. One of the small settlements on the way, known as Wilton, was where he opened fire again. He used his rifle to shoot a mother who was out walking with her adult son. Soon after, he shot that woman's husband as well. They both died.

The route he was taking was completely random. It was hard to predict. It was essential to get as many officers into the area as possible in order to locate this madman and put an end to his rampage. Every available resource was now searching. Cumbria had managed to arm additional officers who were not tasked with armed policing duties that day, creating more patrols that could try to intercept Bird.

Eventually Cumbria Constabulary spotted his taxi, driving in the opposite direction to one of their patrol cars. By then he had shot two more people, killing one. Three ARVs, including Jimmy's vehicle, had been closing in on Bird but were then blocked from entering a tunnel by one of his victims' cars, which had come to a stop in the narrow underpass and needed to be pushed out of the way before the officers could resume their chase. While being filmed by a civilian with their mobile phone, these officers did not – could not – stop to help the victims, instead continuing the hunt. At the time this came in for some criticism, but a review by Simon Chesterman, the ACPO lead for armed policing, found that finding Bird was by far a greater concern and took priority.

More than forty armed officers were now trying to find him. But by now, eleven individuals were dead.

Driving erratically, the gunman stopped at various points to open fire at random passersby, causing more injuries. Outside of a small village called Boot, Bird lost control of his taxi and crashed into oncoming traffic. He abandoned his vehicle and collected his guns, heading into the nearby woodlands. Reporting the crash, some members of the public described what they saw to the police control room, raising hopes that the police helicopter would be able to locate him before armed officers arrived.

The ARVs soon closed in on Bird's taxi, which was missing a front wheel after the crash. Along with two police dog handlers, the armed police officers approached the car and found it abandoned. The decision was taken to conduct an open country search of the woodlands into which Bird had fled. There were now ten armed police officers and two dog handlers.

'The bronze commander[3] shook us out into a line and we headed into the woods,' Jimmy said. 'It would have been an easy place for him to hide... But he was dead by the time we found him. He had eleven bullets left, and nothing for the shotgun.'

Over two hours, Bird had driven 52 miles and shot at least forty-seven rounds of ammunition. His rampage, one of the worst in UK history, cost the lives of twelve people, including Bird, and injury to a further eleven.

---

3 Command in the police service is split into three tiers; Bronze – local, tactical command. Silver – strategic command. Gold – Overall command.

# 25

Operation Accord was finished. The officers involved had returned from Japan and were enjoying post-operation leave. In the meantime, however, we had to start planning for Operation Bluebird, another escort to the Far East. Luckily, many of the details were to be the same. They had to be. There were few facilities in the UK or France with the infrastructure to deal with large nuclear cargo ships, and few people with the skillset to crew them. The same could be said of Japan, where Kobe was typically the receiving port for every shipment. That would make planning a lot easier this time around.

Even without Ed in the office and Matt due to go to Special Branch soon, there was a lot we could get on with. Bones quickly put together a PAR and we set about our work. Once again, I was looking at equipment and clothing alongside training as one of the reserves. I didn't mind that, although all the previous reserve team officers were now on the active team, and I would have liked to be assigned to one of the watches.

'You not putting yourself on the team, Bones?' I asked the skipper as I looked at the list showing which coppers would be on the trip.

'Don't be stupid. There's no way I would go. All that time, at sea, bored out of your skull? Not for me, thanks.'

He had a point. Duty on board was very similar to duty at a power station, except there wasn't even an opportunity to go on patrol – unless a check of the cargo hold counted. By all accounts the most exciting part of the operation was when the Japanese Coast Guard intervention team abseiled from helicopters onto the ships; this was done because the final part of the operation had required the CNC officers to hand over armed security duties once the vessels entered Japanese waters.

Prior to the planning of Operation Bluebird, I had taken the opportunity to fill some gaps in my training that I had not been able to address during the previous year. As a result, I was now trained to the same standard as the other officers on the MEG. Last time, as a last-minute replacement I had missed the initial phase of the training. This time, however, I met up with the team in Fleetwood for a four-day firefighting course that marked the beginning of the pre-embarkation training. I would then be able to go through the entire build-up process and sea trials while assisting with the PAR and operational planning.

One positive piece of feedback we had received was that the presence of a planner – me – on the reserve team allowed queries and requests to be investigated and sorted a lot faster. Officers did not need to wait until office hours to call the planning team; they could get concerns sorted out straight away. I couldn't always fix things instantly, but it was definitely an unintended benefit from the previous operational plan.

The offshore training centre at Fleetwood was designed to educate school leavers who needed to learn everything about the maritime industry, as well as those looking to train in specific areas such as escaping from a helicopter crash while en route to an oil rig, or first-aid training without immediate access to a medical facility. In our case, we were there to practise shipboard firefighting. The instructors, all professional firefighters from the local fire service, were a great bunch. There has traditionally been rivalry between the police and the fire service, though it tends to be good-natured, based on mutual respect and a shared dedication to public service rather than any real animosity.

Putting on our heavy protective clothing I was chatting to the reserve team sergeant, Gary, an older officer who had completed his training in between Accord and Bluebird. Gary had been in the police for his entire career so knew a lot about the job. As a sergeant he was very knowledgeable and good at what he did, but he wanted to get off shift at Sellafield and escape personnel matters for a while, so he joined Escort Ops. He didn't mention any of that in his interview with Wally and Matt, of course. In his late forties, his was an odd career choice as he had a wife and two daughters at home. There were a good number of officers on the MEG who did not have families, as the voyages often went on for as long as military deployments. Spending more time with the family was a common reason for dropping out. But there were always plenty of volunteers ready to fill those vacancies.

'You just did this course didn't you?' I remarked as I fastened my heavy jacket.

'Yup. Same instructors too.'

Some of the newly qualified officers had been training here only a few months ago, and it didn't seem necessary to train them again while the skills were so fresh in their minds. Matt, however, who was still acting inspector, decided that all officers would do all the training together. His reasoning was that it would be a good team-building exercise, especially in the initial stages, and would also eliminate the risk that an officer accidentally missed out on an element of the pre-embarkation.

The instructor started us off easily. They weren't looking to see how fit we were or whether we were suitable for the role; they knew what the MEG was about. Instead, they took us through the basics: rolling out fire hoses so that they could be quickly assembled, activating the hose, and the different ways of delivering water. It was a busy course but probably among the most enjoyable ones I have done. I have never been a particularly competitive person, but when it came time to form small groups to see who could roll out the hose the furthest, or who could suit up the fastest, I suddenly found myself trying to take the top spot. Of course, I never managed. The officers who had only just attended the course as part of their initial training had the edge.

Other aspects of the training included navigating the inside of a ship that was filled with smoke using a portable breathing apparatus. I was surprised to find out how heavy the apparatus was, and just how little air it offered: only around twenty minutes. Navigating the vessel in an oxygen-depleted or smoke-filled environment would need to be efficient and fast, with the objectives and tasks discussed well before entering the area. It struck me during training that the CNC provided all three emergency services while on board the ships: the policing role, of course, but the medical and firefighting roles as well. If something went wrong in the deep sea, there would be no immediate assistance from anywhere else.

The final exercise at the end of the training was a simulated disaster on board. The 'ship' in this case was a maze of shipping containers, stacked on top of and beside one another, that had been set ablaze. There was only one safe entry point, from which the first team went in. With only twenty minutes of air, it was the responsibility of one of the inspectors to keep a chart of who was inside the ship and how long they had been in.

In a narrow and smoke-filled area, it is not as easy as it seems to locate the source of a fire. You can barely see. When you come to a gangway, do you go up or down or straight on? Finding where the blaze is coming from is of course very important, and that information needs to be communicated as soon as possible, but it was also important to consider whether other fires have started owing to the flames making their way into adjoining parts of the ship.

Returning to the start point, the team leader identified what he believed to be the source of the fire, and the second team went in. Successfully extinguishing the blaze, they reported that there were further fires that required immediate action. Further teams went in. When Gary and I went in, the heat was unbelievable. It gave me a lot of respect for the people who choose to fight fires for a living. I couldn't say how hot it was inside there, but it was uncomfortable to the point of distraction. I really had to focus to remember what we had been studying and practising.

With Gary and I on the hose, we began to douse a door with water in preparation for opening it. The hose was set to spray, creating a dense wall of mist in front of us. One of the other team members opened the door and we began to attack the flames in the room. The fire was fierce. It would not go out straight away. But we knew this. This was no log fire or barbeque, and a splash of water was not going to sort it out. We had to attack the flames at their base, methodically reducing the amount of fuel the fire could consume. Eventually we got it under control, and soon enough it was out.

We made our way back to the start point. I took my mask off and looked at the pressure gauge attached to my air tank. It was showing less than 10 per cent. Had we really been in there for almost twenty minutes?

'How much air have you got left mate?' I asked Gary while he was examining his gauge as well.

'Empty,' he laughed. 'I doubt we could have been in there for another two minutes. I'm glad to be out. That was worse than last time. I reckon they upped the ante.'

The whole team looked drained by the exercise. Everyone was sweating and covered in dirt from the burning wood that was used to create the inferno. I could smell the smoke on my clothes; I was looking forward to having the evening off and putting on a clean shirt. We were staying in nearby Blackpool, and one of our team, who had grown up in the town, suggested a night out.

The British police draws its officers from a diverse population. Those who joined the CNC, I found, also came from diverse backgrounds. Pre-embarkation training consisted of four teams: one for the *Heron* and one for the *Pacific Egret* (the replacement for the *Pacific Pintail*), along with a reserve crew for each vessel. The teams covered a broad range of skills, including medical, tactical and technical matters. For that reason, we on the planning team tried to ensure an even balance of officers as well as ensuring friends worked with friends. Inevitably, however, there were occasions where officers did not get along. On a normal policing shift this can easily be managed, but in a small team deployed for extended periods this can cause problems.

Andy was good at his job and had been on the previous operation. He seemed pretty level-headed, but apparently he had not had the best time while on Operation Accord. On our night out in Blackpool he pulled me to the side.

'Princess?'

'What's up, mate?' I asked.

'You need to move me onto the reserve team.'

That surprised me. The teams had already been decided.

'Otherwise I don't think I can go. Not again.' He looked serious. More serious than someone should while enjoying the attractions on Blackpool Pier.

'Alright Andy, I'll look at the teams and speak to the governor in the morning,' I told him, slightly confused but moreover concerned.

'Can't do it,' came the reply from Escort Ops the next day. 'He can either stay or go.'

Matt was not in the mood for any issues in the first week of pre-embarkation training. I was about to act as advocate for Andy when the inspector spoke again.

'Apparently he was hard to work with. Spent most of his time in his cabin playing his PlayStation when he wasn't arguing in the galley. If he's not up for it. He's out.'

I admired Matt's decisiveness but I wondered about Andy. Why was he in Escort Ops if he wasn't up for long voyages with small teams of officers?

'Get him to call me after the weekend,' he finally added before hanging up.

I sought out Andy after the weekend. We were in Wakefield to practise our shooting, and the entire team was in the canteen of the West Yorkshire Police accommodation block.

'The guv wants you to give him a call to talk about changing teams,' I told him. I felt bad about being the messenger. I hoped he would wind his neck in, but I knew that there would be discussions around his suitability.

'Thanks, Princess. Coffee?' He poured me a cup and put it onto my tray. 'I'll call him at lunch.'

I fired off a quick email to warn Matt. He was at work already, training in the HQ gym. He could often be found there. He lived near the constabulary headquarters and didn't need to pay for a private gym membership as a result. He regularly trained with the chief constable as well, which led to some speculation around his appointments in the CNC. I didn't buy it, though. Matt was very good at his job and earned his jobs on merit.

Matt told me that he was in meetings all week, and too busy for random calls.

'Give him the week to think about it,' he concluded. 'I'll speak to him next Monday, but see if you can figure out his position.'

I didn't like the sound of that. I didn't want to push Andy on the way he was feeling, but on the other hand I was part of the planning team. If changes needed to be made, they needed to be made sooner rather than later.

We began our firearms training. By now Jimmy Guns was one of the instructors, along with Chris and Jacko. They had been busy training new members of the MEG but were now committing themselves to the intense build-up for an international shipment. It was good to see Jimmy again. He was a proud new father and showed me more than a few pictures of his daughter. He was a new man. I knew him well enough to know his life had changed for the better. He took us through our G36 training, ensuring that every member of the team was up to standard. He was less casual than he had been, but he retained his sense of humour. Chris, the lead instructor, stood off to one side and said nothing. That was a good sign. There was no way Chris would hold back if something wasn't right.

At the end of the week and the refresher training, I opened an email. It was from Matt. We were about to start our heavy weapons training, this time in Radnor, South Wales. The message was simple and to the point. He must have been busy or working out when he sent it.

'Andy is out. Do you or one of the other reserves want in?'

I wanted to say yes right away, but I decided to first speak with Andy to find out what was happening. Finding his room, I knocked.

'What's going on, mate?' I asked before pushing the door open. Like the rest of us he was packing, ready to go home for the weekend.

'I'm out. It's not for me.'

I paused to take in his decision. Getting into Escort Ops is not easy, after all, and the training takes a long time.

'I miss my boy. I don't want to miss him growing up. He's starting school this year and I think it's for the best. Seeing Jimmy on the training team made me think about getting back on regular shifts.'

I could understand that, despite not having kids myself. In a choice between his boy's first day at school or a voluntary policing role, family won out. I guessed I would have made the same choice if I was in the same position. Soon he would soon be back at his OPU. But his choice had also left me with a decision to make. Should I take his place on the watch, or should I recommend one of the others?

After some consideration, I decided to remain on the reserve team for Bluebird. Instead, I would get myself on the main team for Operation Crown the following spring. I had enjoyed working in planning and alongside the other agencies on the dockside, and I hoped that my experience from the previous op would help Gary in his role as the skipper in charge of gangway security as well. It hadn't been hard to find a volunteer to take Andy's place; one of the newly qualified MEG officers was more than willing to step in.

With that settled, the team travelled to the now extinct county of Radnorshire, where the firearms instructors had arranged for us to do our heavy weapons training. It had been a while since I had used the minigun, so I was keen to practise those skills.

The gunnery range was literally on the side of a mountain, and the targets were on the mountainside opposite; in the valley below, the landowners ran a sheep farm. It had been snowing heavily and the path to the firing point was impassible using normal vehicles. Fortunately, the owners had a whole host of all-terrain vehicles that we could use to get to the top of the track, including a tractor that Chris and Jacko rode in with the guns and mounts.

Once arrived, we set about our tasks: setting up the firing point, loading the magazines, and getting an industrial-sized pot of tea going. It was freezing. I had never been so cold. These were not ideal conditions for training, but things weren't bad enough for us to cancel. At least we could see the targets.

Training felt much better on this op. I didn't feel like an outsider. Whether or not I had proved myself or been accepted was beside the point. My guess was that, by now, I had spoken to most of the escort group in my capacity as operations planner and they knew who I was. Coming off the back of Operation Accord, I had trained alongside most of the team already. I had arranged for their equipment and clothing, as well as taking care of many of the administration matters that found their way to operations planning. I had spoken to most of them via satellite phone during the voyage to Japan and had arranged for their return to the UK. It had been a busy introduction to the world of Escort Ops, but now it felt easier and more manageable.

We conducted a similar training package to before. Our firearms training team was much better resourced now, with four full-time instructors and some volunteers from the firearms training unit. Because of this we were able to run more complex scenarios that reflected the nature of the work we were doing. We carried out armed training as before, but this time without visiting the *Sir Tristram* as we were able to use the *Pintail* while she was being refitted at Barrow-in-Furness. That made planning a lot easier. The teams could simply sleep on their ships before walking to the opposite berth, where she was moored, to begin training. No going up and down the motorway, no sorting out hotels and meals – the ships would provide everything that we needed.

The reserve teams, however, did not have space to stay on board, so it was back to the Abbey House hotel, where I had made private arrangements to stay in my previous lodgings: the Princess Suite.

I met with Gary and the others in the hotel bar that evening.

'So, we have a few visitors tomorrow,' I reminded them.

The plan had been in place for weeks at this point and they knew what was going on. The *Pacific Egret* had not been looked at in detail by our partner agencies or the Royal Navy, specifically those marines who would come to our aid if we lost control of the ship.

'So along with FOST, they're coming up for a half day. Wally is coming up with our admin team and some new regulators from ONR, so that they can have a day out and see what we do in Escort Ops.'

The news wasn't received enthusiastically.

'I know,' I agreed. 'If it's similar to last year, then we will do a run through – the navy will say how shit we are; show us how it's done properly and that's it. Done.'

Nobody likes to be assessed by people they haven't ever met, and it's even harder when there is an audience. For Candice and Amy, however, it would be a great opportunity. They had been working hard to keep the department running smoothly, so I knew they would enjoy a trip out to see the results of their hard work.

# 26

I met the superintendent and his guests at the dockside terminal, where Jimmy Guns and I had spoken to moped rider Mike the previous year. Exchanging greetings, we made our way through the security gates and towards the ships. I recognised the FOST officers from before, so dropped back in order to catch up with them.

'Hello, sirs, how are things?'

It didn't hurt to at least try and find out what I could about the assessments they had planned for us when we conducted our sea trials. Of course they couldn't say, but they let me know that they wouldn't be surprised if a lot of the tasks from last year were used again. That was good enough for me. I made a note to head over to the inspectors when I had a break and let them know. Between them and the team sergeants, it would probably be easy enough to figure out what we might need to do; during this final phase of training, we could focus on similar scenarios.

Concluding the review of the *Egret*, we went across to the *Pintail* in order to conduct some training exercises under the watchful eye of the military. I remembered last time, on the *Sir Tristram*, how we had been a lot slower than the real professionals. But we were very used to this ship. Most of us had worked aboard her at some point, and many of the team had lived on her while she travelled to Japan. If the *Sir Tristram* was an away game, the *Pintail* was the home replay.

The sergeants called the teams together and outlined their plan. In the control centre, the bosses monitored their systems and comms. 'QUICKDRAW, QUICKDRAW, QUICKDRAW' and we were off. Other than some specialist training that is relevant to working in a maritime environment, the AFO skills of the MEG are almost identical to those of the Home Office armed police units. Essentially, the role was the same. We were looking to identify, locate, contain and neutralise. We would also recover control of a part of the ship as well. The room-entry and building-search skills developed and used by the entire UK policing system are therefore very similar to what we were expected to do at sea. The difference came in defending the ship from external threats – using heavy weapons and automatic configurations for our personal weapons.

The teams taking part in the exercises put on a good show. We were fast, but still not as fast as the military. It was another win for the men in black. That isn't to say that we did a bad job, though; far from it. Watching the recorded footage from the ship control room later that day, we were able to review our performance. It was good. The instructors naturally had a few points to highlight, but overall it could not have been much better. FOST, who were with us, agreed.

'Keep it up lads,' they said after they had given us a small debriefing. 'We'll see you again in a couple of weeks.'

True to FOST's word, a lot of the sea trial assessment was very similar to what we had encountered the year before. There was one major difference, though: the killer tomatoes. These huge balloons were designed to be inflated and thrown overboard for use as gunnery targets. Because they are made of rubber, a bullet will pass right through – and although the holes would eventually cause the target to deflate, it took some time. If the damage wasn't too bad, some strong tape could patch up the holes and the tomato could be reused. We had not had these targets before, having made do with wooden cargo pallets left over from the provisions brought aboard by the crew. Being made of wood, these tended to break apart very quickly when hit – especially under a burst of fire from a minigun.

The killer tomatoes also provided a good radar signature, meaning that officers could practise radar skills, which had previously been done using basic software. During the training, an officer in the control centre could identify the target or threat to the ship. This would cause a 'quickdraw', mustering the crew. Once the threat had been confirmed and identified, it was up to the inspectors and sergeants to decide how to engage. It was a much more realistic way to practise as it involved every aspect of the MEG's security training.

On one pass, the inspector determined that all available weapons be brought to engage the target as we drew alongside. This meant not only the heavy weapons, but also those carried by the officers not manning the guns. Waiting on the main deck with the rest of the reserve team, I thumbed the safety of my G36. The first setting was single shot – the second was burst fire. Gary, who was directing our fire, reminded us that the range of the G36 is nowhere near that of a GPMG or minigun. We could hear the bursts of fire from those guns already. The GPMG gave a distinctive and deep rapid staccato, whereas the minigun gave an extremely loud buzzing sound. Our personnel weapons only had a comparatively small magazine as well. Ten bursts of rapid fire was all we could manage before needing to change magazines. By that point, the target would have come into and out of range.

I could see one of the FOST officers in the corner of my eye, his distinctive Royal Navy jacket marking him out. Gary, who was controlling our team, began to fire. Checking our sight pictures, we all began to do the same. For a brief time, the *Egret* was giving a full broadside to a floating red balloon. It seemed ridiculous when we looked back at it later. But when combined with the 30mm cannons mounted on the ship, it went to show just how much firepower we were packing. We arrived back in Barrow docks not long after and concluded the training. In a few days' time, the operational phase of Operation Bluebird would begin.

In a welcome departure from Operation Accord, the initial phase of the operation went almost completely unnoticed. There were no protesters, no journalists and no moped riders with cameras. Gary had paired up with me, and we began our drive down to Portsmouth en route to Cherbourg, where we would once again be helping the French Gendarmerie in securing the port.

Before we could drive onto the ferry, our convoy of cars was pulled aside by a curious Border Force officer. The site of several large cars stuffed full of black bags and blokes had aroused his suspicions. We pulled into a forecourt and got out. Before he could say anything, Gary pulled out his warrant card and police badge. Most of us were still wearing our police firearms coveralls as well, but with civilian jackets over the top.

'State business,' Gary declared before making his way back to the car.

Inwardly I applauded his confidence, but I suspected that the Border Force officer and his gathering colleagues would want a bit more of an explanation. I stepped in.

'We're officers from the Civil Nuclear Constabulary,' I began. I produced my warrant card and badge as well. There was a good chance that this officer had never even heard of the CNC, judging by his expression, so I handed him an Escort Ops business card as well.

'We're doing a job on the other side of the Channel. This is the kit that we need to bring.' I pressed my finger against the glass of the rear window.

Confused, he looked behind him to a more senior colleague, perhaps a supervisor. They were wearing epaulettes with ranks on them, but I was not familiar with the Border Force rank system.

'You're who?' said the supervisor, bluntly.

'Police. Police officers with the Civil Nuclear Constabulary.' I handed him my badge and he took it, looking at it and running his thumb over the raised metal that formed the crown at the top of the badge. By now a few of the other officers had joined me.

'What are you doing?'

I repeated myself. 'We're going to the other side of the Channel to assist with a shipment of nuclear fuel.'

He looked up. I could see what he was thinking. 'What?'

It was not possible to get into the specifics with him – that wasn't allowed – but after some mild negotiation we were released from the forecourt and escorted directly to the ferry. I made a note to make sure that we held a seminar for the Border Force in the future. If they weren't aware of our existence, maybe that should change.

On board the ferry, and after getting into a clean set of clothes, I met the others in one of the restaurants. There wasn't a huge selection of food onboard but there was enough to think about what to have. The crew were French, so it fell to me to speak to the cook manning the servery.

I looked at Gary. 'Can I ask what you want for dinner or is that state business as well?'

He laughed and nudged me as we waited in line. 'I'm not going to hear the end of this, am I?' he asked. It was almost guaranteed that any question he asked in the future would be answered with those two words.

In France, we carried out our tasks in almost the same way as we had before. Once again, the French had put on a big show of force. This time, however, an English-speaking officer had been assigned to each team from the start. These officers were not from the Gendarmerie but the municipal police, having been seconded at the request of the officer planning the French security. That made things a lot smoother.

'The review undertaken after last year highlighted some communication issues,' explained Stessie, the tall, blonde female officer who had been assigned to us. As with the CNC, no doubt there had been debriefings and reviews on the French side after the operation.

Stessie's English was excellent and I told her so: '*Votre anglais est excellent.*' I regretted speaking in French almost immediately, as a barrage of French words came my way faster than I could decipher them. Originally from the Normandy area of France, Stessie had enjoyed visiting the UK and Channel Islands for most of her life, so regularly put her English into practice. It certainly trumped my acquired knowledge from CDs and books and the odd conversation with Candice.

Within a few days the loading phase of the operation had concluded. The *Egret*, which was leading the convoy, made her way out of Cherbourg harbour and began her long voyage. Those of us on the reserve team decided to visit Vierville-sur-Mer, site of the Omaha Beach landing during the Second World War. It was on the way back to the port, and we figured that we had enough time to pay the town a short visit. Walking along the beach, some of the officers talked about a famous movie starring Tom Hanks that featured the assault on Omaha Beach while I spoke to Gary to get his view on how things on Bluebird had gone.

'I loved it,' he said. 'I can't wait to go on a trip.'

We had already looked into some of the aspects of the next operation, Crown, but the planning phase was still in its infancy.

'Wouldn't the wife and kids miss you?' I asked.

'Yeah, no doubt they would. But if I don't do it now, will I ever get the chance again?'

It was a good point. There would certainly have been a lot of people on Omaha Beach who never had the chance to get stuck into a career, start a family or tell customs officials that they were on 'state business'.

# 27

The planning team and senior officers conducted a thorough debriefing after Operation Bluebird and compared it with Operation Accord. What went well? What went wrong? Did this work better, or was it better that way? The details were recorded so that they might be used to make Operation Crown go as well as possible.

One point that was brought up at the end of the meeting was that the police use of firearms was going to change for the entire constabulary, meaning the escort group's training would have to be brought up to date. Although we were a standalone service and not under the auspices of the Home Office, the chief constable, Mr Thompson, had decided that the CNC would be best brought in line with the rest of the country's firearms departments. This was not only to facilitate mutual aid, as highlighted by the Derrick Bird incident, but also to allow the CNC to provide specialist support to other constabularies when the need arose.

For that reason, we were to be licensed by the National Policing Improvement Agency (NPIA), a governmental body that oversaw policing standards in the UK.[1] While our unique aspects, such as

1 The National Policing Improvement Agency (NPIA) was in existence from 2007 to 2013, when it was replaced by the current College of Policing.

automatic and heavy weapons use, would remain 'in house', the rest of the CNC would need to have a review of firearms usage conducted by the NPIA. This was a big deal for us. It meant the REG and MEG when deployed on trans-shipment duties would need to go over their tactics and policies. It would be a major embarrassment if the CNC was to fail in its licensing goals because of the escort department.

Ed, who was now back with us, had further news. The CNC had lost its Investors in People accreditation, much to the consternation of the assistant chief constable, who oversaw personnel and management matters. It didn't surprise me. Candice had been overworked and I knew of several other staff at HQ who were also feeling hard done by. Not a lot was being done to support them, meaning that they had little option but to seek help from outside the service. I remembered the assistant chief constable calling work in the CNC 'a ten-year job at most', and I knew for a fact that the turnover for AFOs was extremely high. My own course had lost over half of its intake within two years. Some stations and OPUs were so understaffed that the cops on duty were given mandatory overtime every week. Maybe parts of the constabulary invested in people, but from what I could see the CNC only invested in a few.

I had been called to an interview with some of the Investing in People assessors when they visited HQ. They needed to speak to a broad range of staff, including the lowest of the low. A PC like me fit that criteria nicely. I told them that, as far as I was concerned, the CNC had been good to me and I personally enjoyed what I was doing.

'And do you have opportunities for advancement?' they asked.

'Yes. I'm learning a lot. I'm not interested in promotion. I like being a constable and I'm not looking for promotion. But the guys I work with, they both got promoted within two years of joining the department.'

I was one of the lucky ones, though. What was it like on a large site like Sellafield or Harwell? What about Dounreay? I had spent time on all the sites and to a certain extent they fulfilled the stereotype of the CNC being nothing more than armed security.

'I think there's some very over cautious management,' I added, 'which I get. But I feel like it's an abundance of caution.'

Pushed for more, I went on. 'It's almost as though someone gets promoted into management and does nothing to benefit their people. Only their own CV. To chase the next promotion. There's no real incentive other than an increase in wages.'

Their pens were hard at work now, so I delivered my conclusion. 'Advancement, as you put it, doesn't need to mean promotion – like the CNC thinks, but just the opportunity to...'

I was finding it hard to finish the sentence. All of a sudden I realised I had been making my own luck and had put myself out there. I was investing in myself. I had one of just two PC roles in escort planning, while the majority were on the gates or on patrol.

'... I don't think there's much opportunity if you aren't prepared to move across the country. That's all. Here at HQ, you can move up, sideways, whatever.'

We had started to plan the next operation to Japan when, on 11 March 2011, the country was hit by a huge earthquake – the biggest in living memory. The subsequent tsunami, which inundated the coastal towns and villages around Fukushima and the north-west, caused over 20,000 deaths as well as damage to the nearby nuclear facility. The engineers in Japan, subsequently dubbed the 'nuclear ninjas', were barely able to keep the emergency contained. Sea defences were supposed to protect the plant's cooling systems from tidal waves, but a 46-foot-high wall of water easily overcame them, compromising the cooling functions, destroying most of the emergency power systems and flooding vast areas of the power station, particularly those built underground.

The disaster sent shockwaves around the world, highlighting the need for nuclear security. If mother nature could cause this much damage without entering the reactors or getting anywhere near the fuel, what could a determined attacker do? Targeting the cooling or other support systems on a site would certainly be easier than breaching the reactor buildings. The CNC had to begin looking with renewed focus at protecting the critical infrastructure of the sites to which they were deployed. In Germany, less than two months after the tsunami, the decision was taken to start to shut down the national nuclear industry in its entirety. Eight of Germany's seventeen nuclear power stations were taken offline almost immediately, and those that remained active were scheduled to be shut down over the coming decades.

Despite its many benefits – near limitless potential for power generation, the creation of highly skilled jobs and secondary industries to support the nuclear sector – the several disasters and many near misses over the decades have caused many to view nuclear energy with some concern. Many people have heard of Chernobyl, Fukushima and Three Mile Island, even though few can point to their locations on a map. It is understandable that people would feel concern not only about the dangers of power generation but also the byproducts. The areas of Chernobyl that can be accessed will not be habitable again until 2065 – seventy-nine years on from the disaster. Those parts buried in concrete – the area where the meltdown occurred – will not be safe to enter for possibly 20,000 years. Scientists just don't have a firm answer.

Longevity aside, the release of excess radiation can have wide-reaching effects. Fallout, or radioactive particles, can easily be swept into the air and travel around the world, moving with clouds and weather systems. This can affect the global food chain and the health of every person on the planet. Concerns around nuclear energy, therefore, are not without merit. Explaining their decision to phase out nuclear power, the German government declared that it wished to 'make Germany safer and avoid additional high-level radioactive waste. *The risks of nuclear power are ultimately unmanageable.*'[2]

With Operation Crown on hold, we could begin to plan an escort to the port of Nordenham in Germany. Surprisingly, we were not picking up fuel that was surplus to requirement now that power stations were closing in the country; on the contrary, we were to drop it off. Eight nuclear fuel rods were due to be delivered from Sellafield to Grohnde, one of the German nuclear plants that remained active, with a subsequent run to deliver more to a different power station.

When carrying out fuel deliveries in and around Europe, PNTL used different ships. Instead of huge vessels like the *Heron*, used for voyages to Japan and America, smaller vessels were used to negotiate European waters. In 2011, a converted military ferry called the *Atlantic Osprey* was allocated for the trip. Already ageing out of service, it was due to be replaced by the *Pintail*. However, as the *Pintail* was in drydock for refitting, the *Osprey* was allocated two final trips before it would be decommissioned and broken up. Being a ferry, however, made the *Osprey* really straightforward to use. Instead of having to load the cargo with a crane and dockside assistance, cargo could just be driven on and off. A maritime escort would be short and simple.

Inspector Mills was chosen to head up the first run, with Inspector Roland and his team on the second. Mills had been on every escort

2 Quote from the German Federal Ministry for the Environment, Nature Conservation, Nuclear Safety and Consumer Protection.

operation since the first run to Japan with the UKAEAC, along with several police veterans such as Sergeant Buckden, who had trained me years before when I was a recruit. He was without a doubt the most experienced inspector available. However, he still tended to rely on his sergeants for advice before making decisions. There was concern growing around his suitability for Escort Ops. Perhaps a temporary inspector would be better.

Inspector Ronald, however, was new to his rank. He had been a PC on Operation Accord but had completed his duties as a sergeant in no time at all, and now he had made inspector in a matter of months. I didn't ask how. I knew he was normally based at Sellafield, where officer turnover is very high. I also knew he was well regarded and respected. He took his work seriously. I also took my work seriously, but I was not interested in promotion. I had been in the job for several years, and many of the people I had worked with had achieved promotion. Gemma, Knoxy and JT were all sergeants. Nix was a detective. My colleagues in ops planning had been promoted. To me, though, working in Escort Ops as a PC was a promotion. At this point, my career was not about rank but role.

I had been assigned to Inspector Mills. Chris and Jacko from firearms training were also aboard, Ed having permitted them to act as part of the crew once pre-embarkation training was completed. I thought it was a great idea. It was only a short trip, so it would not have an impact on any of the ongoing training, and it would allow the firearms instructors to review their training outcomes firsthand. But the team was smaller than for a transpacific operation, and some of the *Osprey*'s onboard facilities were different. The role was the same, however: defend, deny, recover.

As we had already been preparing for Operation Crown, the transition to a European job was quite simple. Bones and Ed had everything in place.

'We need you to do an ammo run. That means you'll miss out on the sea trials,' Bones told me.

BAE Systems, who manufactured our ammunition, had a new batch of 7.62mm ready for us. Unfortunately, it would be at the ammo bunker at the same time that the *Osprey* was due to be at sea with FOST.

'You've already been out once and reserve on another so just join the boat when you've got the bullets,' Bones added.

Ed chipped in. 'It's only a two-day-long sea trial.' He didn't look up from his computer but continued. 'There won't be any live firing either... which is why you need to collect the ammo. FOST were happy with what they saw for Op Bluebird and will allow most of the assessments to carry over.'

Looking at the calendar, I saw that I would be delivering the ammo a few days before the crew was due and the fuel was loaded. That meant I had a four-day weekend.

'No problem,' I told him, simultaneously messaging my girlfriend to suggest a long weekend abroad. Maybe Dublin, or Paris.

I met up with Jimmy at Sellafield and we made our way to the ammunition bunker.

'Never thought I'd be doing this again,' he sighed.

We had both been in the infantry and had already moved a lifetime's worth of ammunition boxes between us.

'Let's get a pub lunch on the way back,' he added.

'We can't mate. Well, you can if you like but I'll have to wait with the wagon. We can't leave the stuff unattended. Boss said so.' Recently, a Glock 17 pistol had gone missing from the new shooting venue at Bisley. During some regularly held training, the ammunition loading point had apparently been left unattended for a few minutes. This meant that someone had enough time to see the firearm, pocket it and walk off. Bisley, being a national shooting venue, is open to the public. The CNC at that time would just hire out one of the ranges, which was little more than a field and a shed, and then conduct firearms training. It was a serious embarrassment for the CNC. The London Olympics were less than twelve months away, and somewhere near London somebody was now equipped with a self-loading pistol. A Freedom of Information request made years later confirmed that the gun has never been found.

'We'll get a Burger Star then,' he said as he stretched into the passenger seat. The journey was going to take quite a while.

With the *Osprey* back from sea trials, there were only two things left to do: load the equipment and load the fuel. The REG was due to escort the rods from Sellafield to Workington aboard a train, where it could be loaded onto the transporter and driven onto the *Osprey*. We would then depart into the Irish Sea and make our way to Europe. Operation Alpina One, as it was known, was therefore split into two elements: the rail escort and the marine escort. Operation Alpina Two was planned for a later date using almost identical details.

Moving fuel from Sellafield to Workington was regular business for the REG, however, and plans had been established way back when the CNC was still the UKAEAC. Nevertheless, Bones and Ed had reviewed the operational order and put the REG onto some build-up training. The marine element of the escort had also been through similar training for the longer voyages, but this time aboard the *Osprey*, a smaller vessel with different armaments and systems.

I was looking forward to being an active part of the operation. It would make a change to planning and logistical work. Gary was aboard along with several of the other newer members of the escort group, who had acted as gangway security and reserves during Operation Bluebird. I had been aboard the *Osprey* before, with Chris, the firearms instructor. Long before the operation had even begun, PNTL, the shipowners, had wanted to install some half-inch steel armour plating around the gunnery positions on the ship. Chris and I had driven over from Sellafield with a GPMG and a few boxes of bullets to test the positions once they had been upgraded. After we had fired off a few hundred rounds and were satisfied that the armour plating did not interfere with security duties, the third officer took us for a tour. He had previously been a deckhand with PNTL, and this was his first executive position in the fleet. He said he recognised us from Operation Accord, but unfortunately I had not really gotten to know many of the crew.

With Sellafield being so close to Workington harbour, the fuel arrived at the dock in no time at all. It was then lifted onto a specially constructed truck driven by two German operators who would take it through the country once it had arrived. When they were aboard, we made them welcome and showed them to their cabins. The *Osprey* looks back-to-front compared to other ships, with the bridge and accommodation at the front and the loading ramp at the back. In order to position itself at the dockside it had needed to 'reverse' in, almost like a lorry, making departure much easier. With a flat bottom, the *Osprey* had a sufficiently shallow draft to let it work in smaller facilities like Workington and many of the Scandinavian and European ports with which PNTL did business. However, this shallow draft made for a bumpy ride when the weather was anything less than ideal.

The trans-shipment point was being patrolled by officers on overtime from Sellafield. The operation was being controlled from B520 at Sellafield, with Bones acting as the radio point of contact again. I feel that he either enjoyed that job or, more likely, wanted to be the first to know any information. It was a nice break for the

Sellafield officers to get away from the gates and internal patrols that filled their days at the massive nuclear site, and as I stood on one of the bridge wings of the *Osprey* I could see them patrolling and occasionally looking back at the ship. I knew roughly what they were doing; I had planned similar operations and taken part in gangway security myself, though never here.

When the fuel was loaded, the *Osprey*'s ramp was fully raised. It was just a matter of time until the captain decided to head out of the mouth of the River Derwent and into the Irish Sea. The operational order that had been put together and agreed to by all parties involved in the operation stated a window for departure, but the captain had the final say. With armed protection all around the ship, the *Osprey*'s defensive armaments were covered with waterproof sheets and not loaded; we would prepare them once we had got into the Irish Sea. As a result, of the escort group only Inspector Mills, Sergeant Gary and a few others manning the surveillance systems had anything to do. I relaxed in my cabin, which was certainly no hotel room, and thought about the upcoming operation.

I had not been on the sea trials as I had been collecting the ammunition for the job, but when I arrived I was told that most of the scenarios had involved dealing with protesters. It made me think about the operational briefing that Inspector Mills had given us once we were all onboard as well. On the German side, we were expecting protests. Greenpeace had let it be known that they intended to protest the arrival of the rods – but our intelligence also suggested that they did not intend to interfere with the delivery. The *Wasserschutzpolizei* and *Bundespolizei*, or water police and federal police, were expecting to deploy 1,300 officers to the operation. That was more officers than the CNC had in total. Once we arrived within German waters, we would be escorted by fast patrol craft so we would have to bring the guns into the ship's armoury and stay below decks. We would then be acting as a security force against anyone who tried boarding the ship.

The engines had been humming gently for quite some time, but they began to shudder as the *Osprey* made its way out of Workington harbour. Using the tannoy system, the captain announced that we were underway. Now that we were entering the Irish Sea, the shoreside policing was over and we reverted to our marine escort role. While one of the watches (or shifts) was on duty the others would be resting, doing administrative tasks, assisting the crew or standing by.

I was on the first watch along with Gary, and I made my way around the gunnery positions to uncover the weapons. Once that was done, I secured the weatherproof sheets in a stowage box nearby and radioed for someone to bring out a box of ammunition. Fixing the ammo into a gun, the other officer and I would test fire, then unload, and put the sheeting back on to protect the weapons from the harsh conditions. We repeated this process for all of the gunnery positions before heading back inside the ship.

The going was rough and the *Osprey* bounced off the waves as it made its way through the Irish Sea. I did not feel well at all. I leaned against a guardrail and watched as waves came over the front of the ship. My eyes felt heavy and I felt unsteady on my feet. *I'm going to be sick*, I thought to myself. I was supposed to be on patrol around the decks anyway, so I 'patrolled' my way back to my cabin, making it just in time. I was hoping that I would feel better soon after being sick, but no such luck. *I can't stay in here being sick*, I thought to myself. I somehow got back to the bridge without throwing up and met Gary.

He took one look at me. 'Are you alright?'

'I'll be alright in a sec,' I lied.

'I don't think so, mate. Go back to your cabin and lie down. I'll get someone else up here.'

I didn't need any encouragement. Feeling like I had been drinking too much, I bounced off the walls as I stumbled back to my cabin and into the bathroom. I felt terrible. It was the first time in my life I had been seasick. The worst part was knowing that I could do nothing about it.

It took me over twenty-four hours to find my sea legs. The officers on patrol would occasionally come into my cabin to check if I was still alive before letting the inspector know that I was still out of the game. Eventually, though still nauseous, I felt able to work. My head was less foggy, and I went onto deck to get some fresh air. Looking around at the horizon, I couldn't see any land. We were well on our way to Germany now. I decided to head to the bridge to see where we were.

I began to feel unwell again the moment I headed inside, so I made my way up using the passages on the side of the ship, where there was fresh air and a horizon to look at. That seemed to work for me. The third officer was on watch.

'Ah, the ship's ghost. Are you feeling better?'

'Not much. Well enough to work though.'

One of the sergeants who was on duty asked me the same thing. He had been on escorts long enough to know that, even though MEG officers joke about seasickness, it's not a laughing matter to those who are suffering from it.

'Chris is down in the armoury cleaning one of the guns. Go and give him a hand. See how you feel.'

I made my way below. Starting to feel queasy, I bolted for one of the bathrooms next to the cargo hold. Chris must have seen me go in, as he was waiting outside for me once I had concluded my business.

'You look like shit, Princess.'

I probably did. I had been lying on my bunk, without showering, shaving or changing my clothes, all the while vomiting hourly.

'Come with me, I'm putting this gun back in its mount,' he said, ushering me up the gangway.

After what felt like an hour but was probably only seconds, we were back on deck, tools in hand, putting one of the guns back into position.

'Feel better now, don't you?'

I did. The fresh air and visible horizon made a huge difference.

'Your watch is on duty after ours. Me and Jacko have been covering for you. You feel up to it?'

I nodded.

'Alright, well if you feel sick again, let Gary know and we'll cover for you again.'

That was good of them.

'We'll talk about repayment options later,' he finally added, winking and guiding me back inside.

I went to the bridge and to speak with Gary.

'You on watch then?' he asked.

I knew how long the duty was supposed to last and looked at my watch. If I kept going off sick I would be in bed all the way to Germany and back, so I decided to force myself.

'Yeah, I'm on watch. Just keep me out of the cargo hold.'

He assigned me some work to do in and around the bridge. It meant that if I began to feel unwell, I could at least head out to a bridge wing and begin to feel better. It also meant that I could see the horizon as I carried out my tasks. The sea was still heavy and the ship was moving around a lot, not helped by the shallow draft of the flat-bottomed *Osprey*. I was still unwell, but not as bad as before. Minutes in, I was already looking forward to getting off watch for a shower and a bite to eat.

My watch concluded, I made my way to the galley and the steward prepared a simple meal for me. With that and a sweet cup of tea, I began to feel human again. Finally. I made my way back to my cabin, all the time checking to make sure I was feeling better and not just imagining it. *I'll go down to the engine room and back*, I thought to myself, wanting to confirm that I was actually getting better. It was a successful trip. *Excellent*, I thought. *I'm cured!* After a shower, I went back to the galley and picked up some of the food that is left out for anyone who wants it: ship's biscuits, packs of crisps and bottles of cola. I took an armful and went back to my cabin to celebrate my newfound health with a slap-up feast.

Soon feeling fresh, I reported for my next shift feeling embarrassed about the previous day and a half. Nobody said anything, but I was worried that it made me look like I wasn't right for the job. I was 'only a planner', after all. Although I had been fully trained for Escort Ops, strictly speaking it was not my role. No doubt it would be brought up in the debrief. Health and fitness always was. I would need to take Ed's advice from Operation Accord and stock up on ginger biscuits and drink lots of coffee.

We carried out our tasks aboard the ship, and I followed our progress on the navigation equipment. Germany was coming into view. We would need to get ready to dock in Nordenham, where the cargo would be offloaded. Once within German waters, the team that was on watch would bring in the defensive armaments and remain out of sight. Sickness aside, it had been an uneventful trip to Germany. The weather had been bad and the third officer, who had come to see me after I was feeling better, brought me a souvenir: a map of the route we had taken. A green line indicated the part of the voyage for which I had been seasick, with a red line marking the rest of the route to our destination. The green line looked pretty long to me, but at least the red line was longer. 'Something to remember your maiden voyage,' he added.

Being escorted into Nordenham by two armed patrol boats, we could also see fast-moving Greenpeace boats in the water as well. Most of them were full of photographers, but one was not. The activists were circling the ship while the German cops were talking to them over loudspeakers. We were in German waters now and had no legal power to do anything as police officers; that would only happen if the activists tried to get aboard. We could see some of them jumping into the water in orange immersion suits, unfurling banners with anti-nuclear slogans on them. Photographers in nearby boats would take pictures and then they would get back into their boats in order to repeat the process from another angle.

It was fair enough, I thought. At least they weren't getting in the way. If they wanted to get cold and wet and take pictures of our ship, that was up to them. I knew that the nuclear industry has always been controversial. We later found out that in total there were only around sixty protesters compared to more than a thousand German police officers.

Looking out of a porthole in the crew lounge where we were waiting, I could see a bright orange boat circling us. We had slowed considerably, and the ship was getting into position to dock. The German vehicle operators, who had been enjoying a pleasant cruise with good food and a lot of card games with the crew, were by now in their truck, waiting to drive off with the help of a heavily armed federal police escort.

Once the ship had shuddered to a halt, an announcement came.

'QUICKDRAW, QUICKDRAW, QUICKDRAW.'

We were wearing general policing equipment only, our firearms having been stowed away as agreed with the German government. We still had batons, pepper spray and handcuffs though. We went to our predetermined intervention points: the cargo area, the engine room, the bridge, the gangways and the deck. What was going on? The cargo ramp remained raised, although I heard the transporter engine coughing into life.

From my position at the rear of the ship, I could see down onto the dockside. There were a handful of protesters outside of the offloading area, and lots of police. It seemed calm and orderly. Why were we at a quickdraw?

Inspector Mills, using the VHF communications system, soon let us know. The speedboat that had been circling us had been waiting for us to stop. As soon as the engines were shut down, a protester jumped into the water and swam a few metres to the front of the *Osprey*. He then climbed up the bulbous bow[3] and sat on it with a protest banner, getting his picture taken by the many photographers that had come to see the protest. Fearing that the man could climb further up the ship, or that other protesters would try something similar on other parts of the ship, Inspector Mills had deployed us as a precaution against anyone trying to come aboard. In fact, the *Osprey* had been boarded before, in a Scandinavian country, though there had been no security detail aboard as the cargo was deemed to be 'low risk'. Regardless, it meant that Greenpeace were more than capable of getting aboard if they wanted to. If that happened, we would need to make sure that they did not get injured in the process – and then arrest them.

Photographs taken, the protesters left the area around our ship, escorted by police boats. The cargo ramp began to lower, and the transporter left the Osprey. No sooner had it left us than the ramp came up, and once again we were heading out to sea for our journey back to Workington.

With no cargo aboard for the return journey, our onboard activities were very different. We no longer had authority to arm the vessel, so all the weapons, including our personal arms, were cleaned and put into cases, ready to be offloaded. We cleaned the areas that we had been occupying and packed away the equipment that we had been using to monitor the environment around us.

With that done, we went to the galley, where Inspector Mills was waiting for us. The stewards had prepared a buffet, and we sat

3 The bulbous bow at the front of a ship is a protruding part of the structure that reduces drag when the ship is moving.

down to enjoy a meal and discuss the voyage. Of course, one of the topics that came up was my 'seasickness'. This soon morphed into me supposedly complaining my mattress was too firm and sulking. Thankfully, the conversation eventually moved away from me and to the voyage.

'Did you see that periscope?' one of our team said. 'I reckon there was a submarine following us the whole time.'

Chris, and Jacko, who were by far the most experienced among us, doubted the validity of this claim.

'No, really. I saw a periscope when I was on watch,' the copper maintained.

Being as certain as he was, I brought it up with FOST the next time I saw them. They told me that they can 'neither confirm nor deny the presence of a submarine during civil shipping operations'.

# 28

Even though an additional shipment of fuel was scheduled to head to Germany, the move was postponed indefinitely. The political climate in Europe was changing rapidly. In the end, the second batch of fuel would remain at Sellafield for a decade before finally finding its way to the continent.

Escort Operations returned to normal running, ensuring that the personnel needed for escorts were trained and that there was enough equipment and ammunition for any spontaneous operations that might emerge. The team had changed, however.

Matt, who had been working as a detective inspector in Special Branch, then went on to become a police officer in Australia. He came back to the UK after twelve months and told me about how he was getting on. He said it was great. He had been assigned to a rural part of Western Australia, and, although he was a lowly constable again, he said it was one of the best moves he ever made.

Bones had decided to head to an OPU to become a shift sergeant. He had recently married and wanted a more predictable lifestyle – one that did not see him regularly going up and down the country. He invited the entire planning team to his wedding, and we were all happy to see him tie the knot.

Candice decided that moving closer to her daughter in the US was the right thing to do, and before we knew it she was gone too.

There was a new chief inspector in the department as well. The tempo of operations over the past few years was so high that it was felt that a more senior officer was required to run the team. Ed, who had applied for the promotion, missed out. Instead, an officer with more administrative experience was appointed.

For my part, I decided that I wanted to head back to Oldbury. Even though I had enjoyed my time on Escort Ops, the new chief inspector was not keen to let planners take part in operations. 'Planners are planners,' he had said as he outlined his intentions for the department. He also wanted to bring in some of the escort team at Harwell to expand the planning department. 'People with more experience should be involved in the planning process,' he reasoned, now that I was the only planner left. I had taken part in four operations by this point, and that felt like quite a lot of experience. It was true, though, that I had not been on any of the deep-sea voyages. Nothing was scheduled for the immediate future other than training and an overhaul of the department, now known as the SEG – the Strategic Escort Group. As Oldbury was looking for PCs to make sideways moves, I decided to head back to where it all began. I didn't really even need to apply. Ed, who had been an excellent boss, sorted it out for me. And so my tour of Escort Operations came to an end.

I arrived back at Oscar Yankee and found that not a lot had changed. The power station had been permitted by ONR to continue generating electricity, and while the reactors provided energy to the National Grid they made around £1 million per day. Understandably, site owner Magnox was keen to try to keep the station open as long as possible.

There were a lot of new faces at the OPU, however, including a tough former French Foreign Legionnaire, a transferee from the Home Office police, and the son of one of the firearms instructors. The station complement of officers had almost doubled. Mo was still in charge and Beegee was now running one of the shifts. I was assigned to Aled's shift, and that was just fine by me. I had stayed in touch with many of the officers at Oldbury and had bumped into them from time to time. It felt good to be back, and I was looking forward to being back on patrol. I would be returning to a more regular pattern of policing, and would hopefully be a better police officer because of the experiences that I had while on Escort Ops.

'I'm pairing you up with Jason,' Aled told me on my first shift back. We had our briefing in the crew room, cups of tea in hand and armed for the night ahead. It was autumn and it was already dark outside. I hadn't done a night shift in years. Jason was a new officer and had only been at the power station for a couple of weeks. He was a big guy, certainly taller and broader than me, and had been in the Royal Navy before joining the CNC.

'You two can do an external to start with,' Aled continued. 'Show him the roads, if you remember them, and then we'll swap over when you get back.'

We headed out to the ARV and secured our weapons in the arms box at the rear of the vehicle.

'Have you just transferred in?' asked Jason.

'Just transferred back, actually. Yeah, I was on Escort Ops for the past few years before this. This was the station I was at when I started out.'

Jason got into the passenger's seat as I got in and started the engine. 'Why did you come back?' he asked.

I gave a brief explanation. 'I was also lucky that there was a space for a PC to make a sideways move,' I added. 'If I had been a skipper I would have had to go to Hinkley or Harwell. Anyway, I like it here. I bet you will too, it's a good unit.'

We made our way out of the power station and into the villages and settlements of South Gloucestershire.

'You need to book us on with the locals, mate,' I told him as we drove down the road, leaving the OPU behind us. 'Our lot already knows we're here, but you need to let Uniform X-ray know as well.'

He thumbed the transmit button on his radio.

'Uniform X-ray, from Oscar Yankee Zero One, over.'

'Oscar Yankee Zero One, go ahead.'

'Uniform X-ray, show Oscar Yankee Zero One as stat two.'

Then came the familiar response.

'All received.'

# Appendix

## LIST OF ABBREVIATIONS

| | |
|---|---|
| **ACPO** | Association of Chief Police Officers |
| **AFO** | Authorised Firearms Officer |
| **ARV** | Armed Response Vehicle |
| **BCU** | Basic Command Unit |
| **BNFL** | British Nuclear Fuel Limited |
| **CBRN** | Chemical, Biological, Radiological and Nuclear |
| **CCC** | Constabulary Control Centre |
| **CFI** | Chief Firearms Instructor |
| **CNC** | Civil Nuclear Constabulary |
| **CNPA** | Civil Nuclear Policing Authority |
| **CTC** | Constabulary Training Centre |
| **CTSU** | Counter Terrorism Search Unit |
| **DET** | Dynamic Entry Team |
| **FATS** | Firearms Training Simulator |
| **FSU** | Firearms Support Unit |
| **FTX** | Final Training Exercise |
| **Glospol** | Gloucestershire Constabulary |
| **GPMG** | General Purpose Machine Gun |
| **HOSDB** | Home Office Scientific and Development Branch |
| **IAEA** | International Atomic Energy Authority |
| **ITC** | Infantry Training Centre |

| | |
|---|---|
| **MEG** | Marine Escort Group |
| **MOD** | Ministry of Defence |
| **MOX** | Mixed Oxide Fuel, a type of nuclear fuel |
| **NATO** | North Atlantic Treaty Organisation |
| **NFI** | National Firearms Instructor |
| **NPIA** | National Police Improvement Agency |
| **NPFTC** | National Police Firearms Training Curriculum |
| **ONR** | Office for Nuclear Regulation |
| **OPU** | Operational Policing Unit (Power Station Unit) |
| **OST/PST** | Officer Safety Training/Personal Safety Training |
| **PAR** | Plan/Action Register |
| **PNC** | Police National Computer |
| **PNTL** | Pacific Nuclear Transport Limited |
| **PTI** | Physical Training Instructor |
| **QRF** | Quick Reaction Force |
| **RAF** | Royal Air Force |
| **REG** | Road Escort Group |
| **RUC** | Royal Ulster Constabulary |
| **SEG** | Strategic Escort Group |
| **SMP** | Sellafield Mox Plant, where MOX fuel is produced and recycled |
| **TRG** | Tactical Response Group |
| **UKAEA** | United Kingdom Atomic Energy Authority |
| **UKAEAC** | United Kingdom Atomic Energy Authority Constabulary |